MW00608388

QUANTUM THEORY OF FIELDS

Gregor Wentzel

Translated by
Charlotte Houtermans *and* J. M. Jauch

DOVER PUBLICATIONS, INC.
Mineola, New York

Bibliographical Note

This Dover edition, first published in 2003, is an unabridged republication of the work originally published in 1949 by Interscience Publishers, Inc., New York. This is an English translation of the work titled *Einführung in die Quantentheorie der Wellenfelder,* published in 1942 by Franz Deuticke, Vienna.

Library of Congress Cataloging-in-Publication Data

Wentzel, Gregor, 1898-
[Einführung in die Quantentheorie der Wellenfelder. English.]
Quantum theory of fields / Gregor Wentzel ; translated by Charlotte Houtermans and J.M. Jauch.
p. cm.
Originally published: New York : Interscience Publishers, 1949.
Includes index.
ISBN 0-486-43245-9 (pbk.)
1. Quantum theory. I. Title.

QC174.12.W4613 2003
530.12—dc22

2003055498

Manufactured in the United States of America
Dover Publications, Inc., 31 East 2nd Street, Mineola, N.Y. 11501

Preface

In research in theoretical physics of the past years, work in the quantum theory of fields has held a significant place. The student who wishes to become acquainted, by means of periodicals, with this branch of study will often find access difficult, even though he may be fully conversant with elementary quantum mechanics. Many of my colleagues who have introduced their pupils to the original literature agree with me in this. This book may help make that access easier. Naturally, I have wondered whether the theory is well enough established to be dealt with in a text-book, but I believe I can ignore such considerations. Certainly the theory has a problematic aspect (see "Self-energy" in the index). If, in the near future, important progress in this theory could be foreseen, we might expect some parts of this book to become rapidly obsolete, yet we can hardly hope for so favorable a development at the present time. While waiting for the liberating new ideas, we must depend on our present theory, and so it seems worth while to make the theory more easily accessible to the interested. Only those who know the theory can understand the problems.

As the title suggests, this book is only an introduction, not an all-inclusive account. The didactic purpose precludes a too systematic approach. It appeared appropriate to start with the "canonical" quantization rules of elementary quantum mechanics (Heisenberg's commutation relations), even though further investigation shows that these rules are too narrow and must be generalized, as, for example, in dealing with particles obeying the Pauli exclusion principle. On the other hand, I have not attempted to deal separately and in detail with the classical theory of wave fields. No doubt, it would have been instructive to illustrate certain deductions in quantum theory by pointing out the corresponding classical considerations; however, since the operator technique of quantum mechanics often simplifies the calculations, I have preferably used the quantum version.

In the first chapter the fundamentals of the theory are dealt with in a general manner, that is, without specification of the field equations or the Lagrange function. The remaining chapters are devoted to particular field types which by means of quantization are associated with particles of various spin, charge, and mass values. No effort was made to attain completeness of treatment; the fields considered may be regarded as typical examples. Naturally, the electromagnetic field and the electron-wave field cannot be omitted. In order to spare the reader unnecessary difficulties, I have deferred many questions, which could have been handled earlier in the general part, to the particular chapters, although at the cost of some repetition. Still, I believe the general part is indispensable, as it is here that the inner homogeneity and consistency of the theory is manifested to some extent. In §4 especially, the proof that the field quantization is Lorentz-invariant is prepared so far that it can be completed for the particular fields without much trouble; at the same time it becomes clear why the "invariant D-function" automatically appears in the relativistic commutation relations. The reader who finds Chapter I difficult in places is urged to study the examples of scalar fields (§6 and §8) as illustrations. The sections in small type are devoted to special questions and applications. The reader may omit these if he wishes.

It is assumed that the reader is acquainted with such fundamentals as are to be found in the current text-book literature. This applies not only to elementary quantum mechanics but especially to Dirac's wave mechanics of spin-electrons (from §17 on). The reader will find the necessary preparation for instance in the article by W. Pauli in Geiger-Scheel's *Handbuch der Physik*, Volume 24, I.

Zurich G. WENTZEL
August, 1942

Preface to the English Edition

I hope the reader will welcome the English translation of my book—initiated by Interscience Publishers—as much as I do. A rough translation of the text was made by Mrs. Charlotte Houtermans, and then revised and corrected scientifically and linguistically by Dr. J. M. Jauch, who also contributed the Appendix. Dr. F. Coester was kind enough to read proof of the galleys and to prepare the index.

The demand for an introduction into the quantum theory of fields seems to be even larger today than some years ago when this book was written. New interest in the subject has been awakened by recent experimental discoveries: the fine structure anomalies of the hydrogen levels, and the correction to the magnetic moment of the electron, which have been interpreted as electromagnetic self-energy terms; the observation of several kinds of mesons and their artificial production, marking a new development that promises a much better understanding of mesons and may eventually lead to an improved meson field theory of nuclear interactions.

As to quantum electrodynamics, it was not possible to incorporate any references to the new theoretical developments into the present edition. We must be content to provide the reader with such basic information as will enable him to follow independently the original literature now appearing in the periodicals. In the chapters on meson theory no changes were necessary except in §15, dealing with the applications to problems of nuclear physics; here it was easy to modernize the text and to adapt it to the present state of knowledge. In various sections throughout the book references to more recent publications have been added. The most significant addition to the original is the Appendix, on the general construction of the energy-momentum tensor according to F. J. Belinfante.

I hope that the book proves useful to many readers in the English-speaking world.

University of Chicago
January, 1949

G. Wentzel

Contents

Preface v
Preface to the English Edition vii

CHAPTER I
General Principles
§1. The Canonical Formalism 1
§2. Conservation Laws for Energy, Momentum and Angular Momentum 8
§3. Complex Field Functions, Conservation of Charge 12
§4. Lorentz Invariance. Time-Dependent Field Operators and Their Commutation Relations 15
§5. Introduction of the Momentum Space 27

CHAPTER II
Scalar Fields
§6. Real Field in Vacuum 29
§7. Real Field with Sources 37
§8. Complex Vacuum Field 48
§9. Complex Field in Interaction with Protons and Neutrons 53
§10. Combined Charged and Neutral Field 63
§11. Charged Particles in an Electromagnetic Field 66

CHAPTER III
The Vector Meson Field
§12. Complex Field in Vacuum 75
§13. Vector Mesons in the Electromagnetic Field 90
§14. Nuclear Interactions 97
§15. Meson Theory and Nuclear Forces 104

CHAPTER IV
Quantum Electrodynamics
§16. The Electromagnetic Field in Vacuum 111
§17. Interaction with Electrons 124
§18. Multiple-Time Formalism 138
§19. Remarks on the Self-Energy Problem 152

CHAPTER V
The Quantization of the Electron Wave Field According to the Exclusion Principle
§20. Force-Free Electrons 167
§21. Electrons in the Electromagnetic Field 188

CHAPTER VI
Supplementary Remarks
§22. Particles with Higher Spin. Spin and Statistics 203
§23. Outlook 212

Appendix I.
The Symmetrization of the Canonical Energy-Momentum Tensor 217

Index 221

QUANTUM THEORY OF FIELDS

Chapter I

General Principles

§ 1. The Canonical Formalism

In classical physics a "field" is described by one or several (real) space-time functions $\psi_\sigma(x, t)$ which satisfy certain partial differential equations, the so-called "field equations." An alternative procedure is to start with a variational principle chosen in such a way that its Euler differential equations are the same field equations. Let L be a function of the $\psi_\sigma(x, t)$ and of their first time and space derivatives:[1]

$$L = L(\psi_1, \nabla\psi_1, \dot{\psi}_1; \psi_2, \nabla\psi_2, \dot{\psi}_2; \ldots). \tag{1.1}$$

By integration over a volume V and a time interval from t' to t'' we form:

$$\int_{t'}^{t''} dt \int_V dx\, L(\psi_1, \ldots) = I.$$

Varying the function ψ_σ for a fixed region of integration:

$$\psi_\sigma(x, t) \rightarrow \psi_\sigma(x, t) + \delta\psi_\sigma(x, t),$$

subject to the restriction that the variations $\delta\psi_\sigma$ vanish at the boundary of the domain of integration (i. e., at the surface of the volume V and for $t = t'$ and $t = t''$), one obtains in the familiar way:

$$\delta I = \int_{t'}^{t''} dt \int_V dx\, \delta L$$

$$= \int_{t'}^{t''} dt \int_V dx \sum_\sigma \left\{ \frac{\partial L}{\partial \psi_\sigma} \delta\psi_\sigma + \sum_k \frac{\partial L}{\partial \frac{\partial \psi_\sigma}{\partial x_k}} \frac{\partial}{\partial x_k} \delta\psi_\sigma + \frac{\partial L}{\partial \dot{\psi}_\sigma} \frac{\partial}{\partial t} \delta\psi_\sigma \right\}$$

$$= \int_{t'}^{t''} dt \int_V dx \sum_\sigma \delta\psi_\sigma \left\{ \frac{\partial L}{\partial \psi_\sigma} - \sum_k \frac{\partial}{\partial x_k} \frac{\partial L}{\partial \frac{\partial \psi_\sigma}{\partial x_k}} - \frac{\partial}{\partial t} \frac{\partial L}{\partial \dot{\psi}_\sigma} \right\}.$$

[1] Partial derivatives with respect to the time will be indicated by dots: $\partial\psi/\partial t = \dot{\psi}$.

We require now that the classical field be determined by the condition that the integral I shall be stationary ($\delta I = 0$) for arbitrary variations $\delta\psi_\sigma$ which satisfy the above-mentioned conditions and for an arbitrary choice of the integration region. From this it follows that for all times and for all positions:

$$\frac{\partial \mathfrak{L}}{\partial \psi_\sigma} - \sum_k \frac{\partial}{\partial x_k} \frac{\partial \mathfrak{L}}{\partial \frac{\partial \psi_\sigma}{\partial x_k}} - \frac{\partial}{\partial t} \frac{\partial \mathfrak{L}}{\partial \dot{\psi}_\sigma} = 0 \quad (\sigma = 1, 2, \ldots). \tag{1.2}$$

These equations are by (1.1) partial differential equations of the second order at most for the field functions ψ_σ; they are the field equations.[1]

This variational principle can be connected with Hamilton's least action principle of classical mechanics. The field variables ψ_σ differ from the coordinates q_j of a mechanical system of particles because the latter are only functions of the time, while the former depend on the position vector x as well. It is, however, possible to establish a correspondence between ψ_σ and the generalized coordinates q_j if we let the discrete index j, which numbers the (finite) degrees of freedom of the system, correspond not only to the discrete index σ but also to the continuous variable x of the position vector. By this procedure a field may be interpreted as a mechanical system of infinitely many degrees of freedom.

[1] One calls

$$\frac{\partial \mathfrak{L}}{\partial \psi_\sigma} - \sum_k \frac{\partial}{\partial x_k} \frac{\partial \mathfrak{L}}{\partial \frac{\partial \psi_\sigma}{\partial x_k}} = \frac{\delta L}{\delta \psi_\sigma}$$

the "functional derivative" of $L = \int dx\, \mathfrak{L}$ with respect to ψ_σ. With this notation, (1.2) reduces to:

$$\frac{\partial}{\partial t} \frac{\partial \mathfrak{L}}{\partial \dot{\psi}_\sigma} = \frac{\delta L}{\delta \psi_\sigma}.$$

The substitution:

$$\mathfrak{L} \to \mathfrak{L} + \sum_k \frac{\partial}{\partial x_k} \Lambda_k (\psi_1, \psi_2, \ldots) + \frac{\partial}{\partial t} \Lambda_0 (\psi_1, \psi_2, \ldots),$$

where $\Lambda_0, \ldots \Lambda_3$ are arbitrary functions of the ψ_σ, leaves the field equations invariant, for the integral:

$$\int_{t'}^{t''} dt \int_V dx \left\{ \sum_k \frac{\partial \Lambda_k}{\partial x_k} + \frac{\partial \Lambda_0}{\partial t} \right\}$$

may be transformed into an integral over the surface of the space-time region and hence its variation vanishes identically. The $\Lambda_0, \ldots \Lambda_3$ may even depend on the derivatives of the ψ_σ, provided $\mathfrak{L}$ remains independent of the second derivatives.

This analogy may be further expounded by subdividing the space into finite cells $\delta x^{(s)}$ which we distinguish by the upper index s; the value of the field function ψ_σ in the cell (s) shall be denoted by $\psi_\sigma^{(s)}(t)$. The space derivatives of ψ_σ will be replaced by the corresponding differences. In this way it is possible to represent the function $\mathfrak{L}$ (1.1), or rather its value $\mathfrak{L}^{(s)}$ in any cell (s), as a function of the generalized coordinates $q_j = \psi_\sigma^{(s')}$ and of the corresponding velocities $\dot{q}_j = \dot{\psi}_\sigma^{(s')}$. Similarly for the sum over all the cells:

$$L = \sum_s \delta x^{(s)} \mathfrak{L}^{(s)}.$$

The variational principle stated above requires then that the time integral:

$$I = \int_{t'}^{t''} dt\, L\,(q_1, q_2, \ldots, \dot{q}_1, \dot{q}_2, \ldots)$$

shall be extremal for the variations $q_j(t) \rightarrow q_j(t) + \delta q_j(t)$ provided δq_j vanishes for $t = t'$ and $t = t''$ as well as in cells (s) outside the region V. Since V is arbitrary, this corresponds precisely to Hamilton's principle in mechanics. The field equations (1.2), which are Euler's differential equations of the variational principle, correspond to Lagrange's equations of motion with the Lagrange function L. In the transition to the limit of infinitesimal cells, one obtains for the Lagrange function the integral extended over all space:

$$L = \int dx\, \mathfrak{L}\,(\psi_1, \nabla \psi_1, \dot{\psi}_1;\ \psi_2, \nabla \psi_2, \dot{\psi}_2;\ \ldots). \tag{1.3}$$

We shall call the integrand $\mathfrak{L}$ "differential Lagrange function."

In order to make the transition to Hamilton's formalism—still in the framework of the classical theory—we must introduce first the momenta $p_j = \partial L/\partial \dot{q}_j$, which are canonically conjugate to the coordinates q_j: with $L = \sum_{s'} \delta x^{(s')} \mathfrak{L}^{(s')}$ the quantity canonically conjugate to $q_j = \psi_\sigma^{(s)}$ becomes:[1]

$$p_j = \delta x^{(s)} \cdot \frac{\partial \mathfrak{L}^{(s)}}{\partial \dot{\psi}_\sigma^{(s)}}$$

In a field theory it is customary to denote the space-time function, obtained from $\mathfrak{L}$ (1.1) by partial differentiation with respect to $\dot{\psi}_\sigma$ (for constant ψ_σ and $\nabla\psi_\sigma$), as the field canonically conjugate to ψ_σ.

[1] $\dot{\psi}_\sigma^{(s)}$ is found in $\mathfrak{L}^{(s)}$ only, and not in the remaining terms of the sum in $L(s' \neq s)$, since according to the definition (1.1) $\mathfrak{L}$ at the point x depends only on the value of $\dot{\psi}_\sigma$ at the same point (and not on $\nabla\dot{\psi}_\sigma$ also).

(1.4) $$\pi_\sigma = \frac{\partial \mathbf{L}}{\partial \dot{\psi}_\sigma}.$$

Hence, if its value in the cell (s) is $\pi_\sigma^{(s)}$:

$$p_j = \delta x^{(s)} . \pi_\sigma^{(s)}.$$

Hamilton's function is then obtained by:

$$H = \sum_j p_j \dot{q}_j - L = \sum_s \delta x^{(s)} \left\{ \sum_\sigma \pi_\sigma^{(s)} \dot{\psi}_\sigma^{(s)} - \mathbf{L}^{(s)} \right\},$$

H is to be considered as a function of the q_j and p_j or of the ${}_\sigma\psi^{(s)}$ and $\pi_\sigma^{(s)}$. In the limit of the continuum we obtain:

(1.5) $$H = \int dx\, \mathbf{H}, \qquad \mathbf{H} = \sum_\sigma \pi_\sigma \dot{\psi}_\sigma - \mathbf{L}.$$

By eliminating $\dot{\psi}_\sigma$ with the help of (1.4) we may consider the "differential Hamiltonian function" $\mathbf{H}$ as a function of the ψ_σ, $\nabla\psi_\sigma$ and π_σ:

(1.6) $$\mathbf{H} = \mathbf{H}(\psi_1, \nabla \psi_1, \pi_1; \psi_2, \nabla \psi_2, \pi_2; \ldots).$$

Introducing the Hamiltonian H makes it possible to replace the field equations (1.2) by the "canonical field equations" which correspond to the canonical (Hamilton) equations of motion in classical particle mechanics:[1]

$$\dot{q}_j = \frac{\partial H}{\partial p_j}, \quad \dot{p}_j = -\frac{\partial H}{\partial q_j}$$

We shall, however, not elaborate on this point here, since it is not essential for obtaining the corresponding equations in quantum theory.

Since our classical ψ-field is, as we have seen, equivalent to a mechanical system of particles, with infinitely many degrees of freedom, it is natural to carry out its quantization according to the standard rules of quantum mechanics. The transition from classical to quantum mechanics can be effected by replacing the canonical variables q_j, p_j, by Hermitian operators which satisfy the commutation rules:[2]

$$[q_j, q_{j'}] = [p_j, p_{j'}] = 0, \quad [p_j, q_{j'}] = \frac{h}{i} \delta_{jj'}$$

[1] More details in: Heisenberg and Pauli, *Z. Phys.* 56, 1, 1929, §1.

[2] Our notations are the usual ones: $[a, b] = ab - ba$, h = Planck's constant divided by 2π, i = imaginary unit, $\delta_{jj} = 1$, $\delta_{jj'} = 0$ for $j \neq j'$.

The mechanical properties of the system are determined by its Hamiltonian which is formally taken over from the classical theory, but reinterpreted as Hermitian operator. In a similar way we may consider the properties of a quantized field as defined by its Hamiltonian $H = \int dx\, \mathsf{H}$ or also by its Lagrangian $L = \int dx\, \mathsf{L}$, from which H may be obtained according to (1.4) and (1.5). The $\psi_\sigma(x)$ and $\pi_\sigma(x)$ in H (1.6) stand now for Hermitian operators, with commutation rules which result from those of $q_i = \psi_\sigma^{(s)}$, $p_i = \delta x^{(s)}\, \pi_\sigma^{(s)}$ by the transition to the continuum. This procedure is characteristic of the so-called "canonical field quantization" as it was first formulated for general fields by Heisenberg and Pauli.[1]

Accordingly we postulate the following equations:

$$\left[\psi_\sigma^{(s)}, \psi_{\sigma'}^{(s')}\right] = \left[\pi_\sigma^{(s)}, \pi_{\sigma'}^{(s')}\right] = 0, \qquad \left[\pi_\sigma^{(s)}, \psi_{\sigma'}^{(s')}\right] = \frac{h}{i}\, \delta_{\sigma\sigma'} \cdot \frac{\delta_{ss'}}{\delta x^{(s)}}.$$

Writing $\psi_\sigma(x)$, $\pi_\sigma(x)$, instead of $\psi_\sigma^{(s)}$, $\pi_\sigma^{(s)}$, we find for the commutation rules:

$$(1.7)\quad [\psi_\sigma(x), \psi_{\sigma'}(x')] = [\pi_\sigma(x), \pi_{\sigma'}(x')] = 0, \quad [\pi_\sigma(x), \psi_{\sigma'}(x')] = \frac{h}{i}\, \delta_{\sigma\sigma'} \cdot \delta(x, x');$$

Here $\delta(x, x')$ stands for a function the value of which is $(\delta x^{(s)})^{-1}$ or 0, according to whether the points x and x' lie in the same or in different cells. Integrating for fixed x' with respect to x, we obtain $\int dx\, \delta(x, x') = 1$. Furthermore, $\int dx f(x)\, \delta(x, x')$ is equal to the average value of the function $f(x)$ taken over the cell (s') in which x' is situated. In the limit of the continuum (cell volume $\delta x^{(s)} \to 0$) $\delta(x, x')$ goes over into the (three-dimensional) Dirac δ-function:

$$(1.8)\quad \delta(x, x') \to \delta(x - x') = \begin{cases} 0 \text{ for } x \neq x', \\ \infty \text{ for } x = x', \text{ in such a way that} \end{cases}$$

$$\int dx\, \delta(x - x') = 1.$$

This limiting process has meaning only if $\delta(x, x')$ appears in the integrand of a space integral.

$$(1.9)\quad \int dx\, f(x)\, \delta(x, x') \to \int dx\, f(x)\, \delta(x - x') = f(x').$$

It has become customary to write directly $\delta(x - x')$, instead of $\delta(x, x')$, and we shall adopt the same notation in the following, since there is no doubt as to its meaning.

[1] Cited above.

In the quantum theory of fields ψ_σ, π_σ represent operators which depend on the coordinates x of the position as parameters[1] and satisfy the commutation rules (1.7, .8) Later on we shall introduce particular representations for these operators when we discuss special fields. It should be emphasized that for the time being we consider the operators $\psi_\sigma(x)$, $\pi_\sigma(x)$ as time-independent, the construction of time-dependent operators being reserved for a later discussion in §4 in connection with the question of relativistic invariance. Together with ψ_σ and π_σ, $\boldsymbol{H}$, too, is a space-dependent operator and the integral Hamiltonian $H = \int dx\, \boldsymbol{H}$ is then an operator independent of position. It may be necessary to arrange non-commuting factors in a suitable way so as to make H a Hermitian operator (symmetrization).

At this point a particular case should be mentioned which plays a certain role in the quantum theory of fields. It happens with certain Lagrangians that some field equations do not contain the second derivatives $\ddot{\psi}_\sigma$ with respect to time; these equations then represent "subsidiary conditions" which establish a relationship between the variables ψ_σ and $\dot{\psi}_\sigma$.[2] In this case the variables ψ_σ, π_σ are no longer independent of each other, and it follows that the commutation rules (1.7) lead to contradictions. A case like this will be encountered in §§12 and 16 where one of the $\dot{\psi}_\sigma$ does not occur in $\boldsymbol{L}$ and consequently the corresponding $\pi_\sigma = \partial \boldsymbol{L}/\partial \dot{\psi}_\sigma$ vanishes identically. The commutation rule $[\pi_\sigma(x), \psi_\sigma(x')] = -i\,h\,\delta(x - x')$ is then obviously incorrect. We will see later how the quantization can be carried through in such a case. One possibility is to eliminate the redundant field components and to postulate the commutation rules (1.7) for the remaining independent variables (cf. §12, page 77). By this elimination procedure the Hamiltonian may become dependent on the space derivatives of the π_σ so that we have instead of (1.6):

$$\boldsymbol{H} = \boldsymbol{H}(\psi_1, \nabla \psi_1, \pi_1, \nabla \pi_1; \ldots). \tag{1.10}$$

---

[1] x itself is no operator, but a "c-number."

[2] If the Lagrangian has, for instance, the form:

$$\boldsymbol{L} = \dot{\psi}_1 \cdot F + G,$$

where F is independent of all $\dot{\psi}_\sigma$ and G of $\dot{\psi}_1$, then the equation (1.2) with $\sigma = 1$ has evidently this above-mentioned character of a subsidiary condition. For then:

$$\pi_1 = \frac{\partial \boldsymbol{L}}{\partial \dot{\psi}_1} = F,$$

i.e., π_1 depends only upon the ψ_σ and their space derivatives. If the operators ψ_σ all commute with each other, it follows that π_1 also commutes with all ψ_σ, in contradiction to (1.7).

The following considerations hold also for this more general type of Hamiltonians.

In quantum mechanics, as is well known, the canonical equations of motion are valid as operator equations on account of the commutation rules of the q_j and p_j:

$$\dot{q}_j \equiv \frac{i}{h}[H, q_j] = \frac{\partial H}{\partial p_j}, \quad \dot{p}_j \equiv \frac{i}{h}[H, p_j] = -\frac{\partial H}{\partial q_j}.$$

In general the time derivative of any function φ of the q_j and p_j not depending on time explicitly, may be expressed by:[1]

$$\dot{\varphi} \equiv \frac{i}{h}[H, \varphi] \tag{1.11}$$

Since we have taken over the commutation rules of particle mechanics, the same result holds for the field theory: the time derivative of any field quantity, i.e., any functional of the $\psi_\sigma(x)$ and $\pi_\sigma(x)$ and their space derivatives depending not explicitly on time, satisfies an equation (1.11). In particular the operators $\dot{\psi}_\sigma$ and $\dot{\pi}_\sigma$ are defined by:

$$\dot{\psi}_\sigma(x) \equiv \frac{i}{h}[H, \psi_\sigma(x)], \quad \dot{\pi}_\sigma(x) \equiv \frac{i}{h}[H, \pi_\sigma(x)] \tag{1.12}$$

The evaluation of these commutators with the help of the commutation rules (1.7,.8) leads to operator equations which are formally equivalent with the field equations (1.2) in the same way as the canonical equations of motion in particle mechanics ($q_j = \partial H/\partial p_j$, $p_j = -\partial H/\partial q_j$) are equivalent to Lagrange's equations. We forego here a general proof.[2] In the following application of the theory to special types of fields, we shall verify this equivalence in each case. On account of their analogy to the canonical equations of motion, we shall refer

[1] To clarify this it is to be remembered that φ signifies an operator which does not depend on time explicitly. The operator $\dot{\varphi}$ as defined by (1.11) is therefore not simply the partial derivative of φ with respect to time. It derives its significance, however, from the well-known theorem in quantum mechanics that the expectation value of $\dot{\varphi}$ is equal to the time derivative of the expectation value of φ, thus:

$$\bar{\dot{\varphi}} \equiv \frac{i}{h}\overline{[H, \varphi]} = \frac{d\bar{\varphi}}{dt}.$$

[2] Cf. footnote 1, page 2, Heisenberg and Pauli. The commutators (1.12), are equal to the functional derivatives of the Hamiltonian H with respect to π_σ and $-\psi_\sigma$, respectively.

$$\frac{i}{h}[H, \psi_\sigma] = \frac{\delta H}{\delta \pi_\sigma}, \quad \frac{i}{h}[H, \pi_\sigma] = -\frac{\delta H}{\delta \psi_\sigma}.$$

[Cf. footnote 1, page o; from it results the stipulated equivalence with (1.2)].

to the operator equations which stem from the definitions (1.12) as "canonical field equations."

All questions regarding the stationary states of the system, the eigen values of H or any other field quantities, can be answered with the well-known methods of quantum mechanics, notwithstanding the fact that we are dealing with a system of infinitely many degrees of freedom. Examples will follow. Before we go into this, some questions of a more general character shall be discussed which have no analogy in particle mechanics.

§ 2. Conservation Laws for Energy, Momentum, and Angular Momentum

We return to the classical (non-quantized) theory. The conservation of energy is expressed by the equation $dH/dt = 0$, provided that the Lagrangian does not depend explicitly on time. It is to be expected that the application of this conservation law to the integral Hamiltonian function (1.5), which represents the total energy of the field, leads to the interpretation of the differential Hamiltonian function $\boldsymbol{H}$ as an energy density, for which a continuity equation holds.

$$\frac{\partial \boldsymbol{H}}{\partial t} + \nabla \cdot \boldsymbol{S} = 0. \tag{2.1}$$

One obtains in fact from (1.4,.5):

$$\frac{\partial \boldsymbol{H}}{\partial t} = \sum_\sigma \left\{ \dot{\psi}_\sigma \frac{\partial}{\partial t} \frac{\partial \boldsymbol{L}}{\partial \dot{\psi}_\sigma} + \ddot{\psi}_\sigma \frac{\partial \boldsymbol{L}}{\partial \dot{\psi}_\sigma} - \left(\dot{\psi}_\sigma \frac{\partial \boldsymbol{L}}{\partial \psi_\sigma} + \ddot{\psi}_\sigma \frac{\partial \boldsymbol{L}}{\partial \dot{\psi}_\sigma} + \sum_k \frac{\partial \dot{\psi}_\sigma}{\partial x_k} \frac{\partial \boldsymbol{L}}{\partial \frac{\partial \psi_\sigma}{\partial x_k}} \right) \right\},$$

and using the field equations (1.2):

$$\frac{\partial \boldsymbol{H}}{\partial t} = -\sum_\sigma \sum_k \left\{ \dot{\psi}_\sigma \frac{\partial}{\partial x_k} \frac{\partial \boldsymbol{L}}{\partial \frac{\partial \psi_\sigma}{\partial x_k}} + \frac{\partial \dot{\psi}_\sigma}{\partial x_k} \frac{\partial \boldsymbol{L}}{\partial \frac{\partial \psi_\sigma}{\partial x_k}} \right\} = -\sum_k \frac{\partial}{\partial x_k} \sum_\sigma \dot{\psi}_\sigma \frac{\partial \boldsymbol{L}}{\partial \frac{\partial \psi_\sigma}{\partial x_k}};$$

hence the continuity equation (2.1) is satisfied by:

$$S_k = \sum_\sigma \dot{\psi}_\sigma \frac{\partial L}{\partial \dfrac{\partial \psi_\sigma}{\partial x_k}}. \tag{2.2}$$

This definition of the energy current density is however not unique, since any source-free field could be added to $\mathbf{S}$.

Furthermore we will try to define the momentum of the field:

$$G = \int dx \, \mathbf{G} \tag{2.3}$$

so that momentum is conserved:

$$\frac{\partial G_k}{\partial t} + \sum_j \frac{\partial T_{jk}}{\partial x_j} = 0, \tag{2.4}$$

Here T is a stress tensor. With the notation:

$$x_4 = i\,c\,t, \tag{2.5}$$

$$T_{44} = -H, \quad T_{k4} = \frac{i}{c} S_k, \quad T_{4k} = i\,c\,G_k \quad (k = 1, 2, 3) \tag{2.6}$$

one can summarize the conservation laws (2.1,.4)

$$\sum_{\mu=1}^{4} \frac{\partial T_{\mu\nu}}{\partial x_\mu} = 0 \quad (\nu = 1 \,..\, 4). \tag{2.7}$$

An expression for the energy-momentum tensor T, which agrees in the 4,4- and k,4-components with (1.5) and (2.2), is:

$$T_{\mu\nu} = -\sum_\sigma \frac{\partial \psi_\sigma}{\partial x_\nu} \frac{\partial L}{\partial \dfrac{\partial \psi_\sigma}{\partial x_\mu}} + L \, \delta_{\mu\nu}. \tag{2.8}$$

This expression also satisfies the conservation equations (2.7), provided that L does not depend explicitly on the x_ν ($\nu = 1 \ldots 4$), for a short calculation gives:

$$\sum_{\mu=1}^{4} \frac{\partial T_{\mu\nu}}{\partial x_\mu} = -\sum_\sigma \frac{\partial \psi_\sigma}{\partial x_\nu} \left\{ \sum_{\mu=1}^{4} \frac{\partial}{\partial x_\mu} \frac{\partial L}{\partial \dfrac{\partial \psi_\sigma}{\partial x_\mu}} - \frac{\partial L}{\partial \psi_\sigma} \right\},$$

and this vanishes on account of the field equations (1.2), which with the notation

(2.5) may be written:

$$\sum_{\mu=1}^{4} \frac{\partial}{\partial x_\mu} \frac{\partial L}{\partial \frac{\partial \psi_\sigma}{\partial x_\mu}} = \frac{\partial L}{\partial \psi_\sigma}. \tag{2.9}$$

Thus conservation of momentum holds, if the momentum density G of the field is defined as follows:

$$G_k = \frac{1}{ic} T_{4k} = -\sum_\sigma \frac{\partial \psi_\sigma}{\partial x_k} \frac{\partial L}{\partial \dot{\psi}_\sigma} = -\sum_\sigma \frac{\partial \psi_\sigma}{\partial x_k} \pi_\sigma. \tag{2.10}$$

Of course this does not mean that (2.8) is the only expression which satisfies the conservation equations (2.7). The tensor (2.8) is called the "canonical energy-momentum tensor."

One should furthermore postulate that the energy-momentum tensor be a symmetrical tensor: $T_{\mu\nu} = T_{\nu\mu}$, for, as is well known, this is the condition that we have, in addition to conservation of energy and momentum, also conservation of angular momentum.[1] Whether the canonical tensor satisfies this condition depends on the nature of the Lagrangian L. In the following examples (§§5–10) we shall deal with the simplest fields for which the conditions of symmetry will be fulfilled. If the conditions are not satisfied, the possibility exists to complete T to a symmetrical tensor by adding a divergence-free tensor $T'\left(\sum_\mu \partial T'_{\mu\nu}/\partial x_\mu = 0\right)$; for this case, too, we shall give examples (§§11 ff.).[2]

In the derivation of the continuity equations (2.1 and .4, resp. .7) we have explicitly assumed that the Lagrangian L depends only indirectly on x and t

[1] If one defines an angular momentum density:

$$M_{\mu\nu\nu'} = T_{\mu\nu} x_{\nu'} - T_{\mu\nu'} x_\nu,$$

its divergence according to (2.7) becomes:

$$\sum_\mu \frac{\partial M_{\mu\nu\nu'}}{\partial x_\mu} = T_{\nu'\nu} - T_{\nu\nu'},$$

Hence this equals zero, if T is symmetrical. From this it follows that the skew-symmetrical tensor $\int dx M_{4\nu\nu'}$ is constant in time. In particular, the components of the angular momentum are constant:

$$\frac{1}{ic} \int dx\, M_{4kk'} = \int dx\, (G_k x_{k'} - G_{k'} x_k).$$

[2] Belinfante (*Physica 6*, 887, 1939) and Rosenfeld (*Mém. de l'acad. roy. de Belgique XVIII*, No. 6, 1940) give a general rule for the construction of T'. Rosenfeld starts from the general theory of relativity which, as is well known, connects the energy-momentum tensor with the gravitational field, whereas Belinfante uses only the postulate of invariance of the special theory of relativity; cf. Appendix I.

through its dependence on the ψ_σ and their derivatives. If L depends, in addition to these, explicitly on x, t (as, for instance, in the case of the Schrödinger wave function of a particle in a given force field), one finds instead of (2.7) for the canonical tensor (2.8), with regard to (2.9):

$$\sum_\mu \frac{\partial T_{\mu\nu}}{\partial x_\mu} = \frac{\partial L}{\partial (x_\nu)}, \tag{2.11}$$

The right hand side denotes the derivative of L with respect to x_ν, for constant ψ_σ, $\dot{\psi}_\sigma$ and $\nabla\psi_\sigma$. The equation (2.11) expresses the fact that at space-time points, where $\partial L/\partial\,(x_\nu)$ does not vanish, there takes place an exchange of energy and momentum of the field with the external systems, which interact with the ψ-field.

In order to carry out the transition to quantum theory, one must represent the tensor components $T_{\mu\nu}$ (in the same way as earlier $H = -T_{44}$) as functions of the variables ψ_σ, $\nabla\psi_\sigma$, and π_σ with the help of (1.4). These variables are then interpreted as operators with the commutation rules (1.7). The operators, which in the classical theory correspond to real field quantities (S_k, G_k, T_{kl}), must be made Hermitian by an appropriate arrangement of non-commuting factors. The conservation of the total momentum G (we assume again that $\partial L/\partial(x_\nu) = 0$) is revealed by the fact that the operator $G = \int dx\, G$ commutes with the Hamiltonian operator H: $[H, G] = 0$.[1] In order to test the validity of the differential conservation laws (2.1 and 4) in the quantized theory, one constructs the operators:

$$\dot{H}(x) = \frac{i}{h}[H, H(x)], \qquad \dot{G}(x) = \frac{i}{h}[H, G(x)]; \tag{2.12}$$

which correspond to the classical quantities $\partial H/\partial t$ and $\partial G/\partial t$. These operators can be represented as divergences:

$$\dot{H} = -\nabla\cdot S, \qquad \dot{G}_k = -\sum_j \frac{\partial T_{jk}}{\partial x_j}. \tag{2.13}$$

since their space integrals

$$\dot{H} = \int dx\, \dot{H}(x), \qquad \dot{G} = \int dx\, \dot{G}(x)$$

[1] The validity of the conservation laws (as in classical mechanics) is known to be connected with certain invariance properties of the Hamiltonian. For instance, the law of the conservation of momentum is connected with the invariance of H with respect to translations parallel to the coordinate axes. This invariance which exists in the case $\delta L/\delta(x_\nu) = 0$, can be used directly to prove generally the commutability of the momentum operator G (2.3, 10) with H. Cf. Heisenberg and Pauli, *Z. Phys.* 59, 168, 1930, §1.

vanish on account of $[H, H] = 0$, $[H, G] = 0$. The operator equations (2.13) turn out to be entirely analogous to the classical equations (2.1,.4) for any special choice of the Lagrangian. The tensor components $T_{\mu\nu}$ [cf. (2.6)], considered as operators, satisfy therefore these same equations, provided the factors are written in suitable order.

§ 3. Complex Field Functions. Conservation of Charge

So far our considerations have referred to real field functions ψ_σ. In nature, however, complex fields are of importance too. An example is the de Broglie-Schrödinger wave function of an electron. One can obviously decompose every complex field function into a real and an imaginary part, represent L as a function of these real field functions ψ_σ, and apply the formalism developed so far. Sometimes it is, however, formally less complicated to work directly with the complex wave functions. We shall explain briefly how the formalism must be modified in this case, using for simplicity the example of a field with only one single (complex) component ψ.

Let:

$$\psi = \frac{1}{\sqrt{2}} (\psi_1 + i \psi_2), \quad \psi^* = \frac{1}{\sqrt{2}} (\psi_1 - i \psi_2), \tag{3.1}$$

where ψ_1, ψ_2 are real functions ($\psi_\sigma = \psi_\sigma^*$). We represent L as function of ψ and ψ^* and their derivatives, rather than as function of ψ_1, ψ_2 and their derivatives:

$$L = L(\psi, \nabla \psi, \dot{\psi}; \ \psi^*, \nabla \psi^*, \dot{\psi}^*). \tag{3.2}$$

In order to obtain the field equations from the variation principle ($\delta I = 0$), one can vary the complex functions $\psi(x, t)$ and $\psi^*(x,t)$ independently, instead of $\psi_1(x,t)$ and $\psi_2(x,t)$, since in this way one obtains the same manifold of varied fields:

$$\psi(x, t) \rightarrow \psi(x, t) + \delta\psi(x, t), \quad \psi^*(x, t) \rightarrow \psi^*(x, t) + \delta\psi^*(x, t).$$

The variation principle now yields, in a similar way as before, the field equations ($x_4 = i c t$, $\mu = 1 \ldots 4$):

$$\sum_\mu \frac{\partial}{\partial x_\mu} \frac{\partial L}{\partial \frac{\partial \psi}{\partial x_\mu}} = \frac{\partial L}{\partial \psi}, \quad \sum_\mu \frac{\partial}{\partial x_\mu} \frac{\partial L}{\partial \frac{\partial \psi^*}{\partial x_\mu}} = \frac{\partial L}{\partial \psi^*}. \tag{3.3}$$

The partial derivatives of L derive their significance from the fact that L is to be considered as a function of the ψ, ψ^* and their derivatives, according to (3.2).

For instance, $\partial L/\partial\psi$ is the derivative with respect to ψ with constant $\nabla\psi$, $\dot{\psi}$, ψ^*, $\nabla\psi^*$, $\dot{\psi}^*$. The equations (3.3) are of course equivalent to the equations (1.2) for $\sigma = 1, 2$.

The fields π, π^*, canonically conjugate to the complex fields ψ, ψ^*, may then be defined as follows:

$$\pi = \frac{\partial L}{\partial \dot{\psi}}, \quad \pi^* = \frac{\partial L}{\partial \dot{\psi}^*}. \tag{3.4}$$

We shall express π, π^* by the real fields π_1, π_2 which are canonically conjugate to ψ_1, ψ_2:

$$\pi_\sigma = \frac{\partial L}{\partial \dot{\psi}_\sigma} = \frac{\partial L}{\partial \dot{\psi}}\frac{\partial \dot{\psi}}{\partial \dot{\psi}_\sigma} + \frac{\partial L}{\partial \dot{\psi}^*}\frac{\partial \dot{\psi}^*}{\partial \dot{\psi}_\sigma} \qquad (\sigma = 1, 2);$$

One obtains with (3.4) and (3.1):

$$\pi_1 = \frac{1}{\sqrt{2}}(\pi + \pi^*), \qquad \pi_2 = \frac{i}{\sqrt{2}}(\pi - \pi^*),$$

or:

$$\pi = \frac{1}{\sqrt{2}}(\pi_1 - i\,\pi_2), \quad \pi^* = \frac{1}{\sqrt{2}}(\pi_1 + i\,\pi_2). \tag{3.5}$$

Since $\pi\,\dot{\psi} + \pi^*\,\dot{\psi}^* = \pi_1\,\dot{\psi}_1 + \pi_2\,\dot{\psi}_2$, it follows for the energy density according to (1.5):

$$H = \pi\,\dot{\psi} + \pi^*\,\dot{\psi}^* - L, \tag{3.6}$$

and analogously for the momentum density (2.10):

$$G = -(\pi\,\nabla\,\psi + \pi^*\,\nabla\,\psi^*). \tag{3.7}$$

It can easily be verified generally that for all components of the tensor (2.8) we have:

$$T_{\mu\nu} = -\left(\frac{\partial L}{\partial\dfrac{\partial\psi}{\partial x_\mu}}\frac{\partial\psi}{\partial x_\nu} + \frac{\partial L}{\partial\dfrac{\partial\psi^*}{\partial x_\mu}}\frac{\partial\psi^*}{\partial x_\nu}\right) + L\,\delta_{\mu\nu}. \tag{3.8}$$

In quantum theory the ψ, ψ^*, π, π^* as well as the ψ_σ, π_σ, become space-dependent operators. The operators ψ_σ, π_σ, which correspond to the real field functions, are Hermitian ($\psi_\sigma = \psi_\sigma^*$, $\pi_\sigma = \pi_\sigma^*$), which is by no means true of the operators (3.1 and .5), since they contain the imaginary unit i. ψ^* and π^* are Hermitian conjugate to ψ and π respectively. Their commutation rules follow directly from these for ψ_σ, π_σ, One finds from (3.1, .5) and (1.7):

$$(3.9)\quad \begin{cases} [\pi(x), \psi(x')] = [\pi^*(x), \psi^*(x')] = \frac{h}{i}\,\delta(x - x'), \\ [\psi(x), \psi(x')] = [\psi(x), \psi^*(x')] = [\psi^*(x), \psi^*(x')] \\ = [\pi(x), \pi(x')] = [\pi(x), \pi^*(x')] = [\pi^*(x), \pi^*(x')] \\ = [\pi(x), \psi^*(x')] = [\pi^*(x), \psi(x')] = 0. \end{cases}$$

The only two pairs of variables which do not commute are thus π, ψ and π^*, ψ^*. This justifies their notation as canonically conjugate pairs: the formulas (3.1 and .5) represent a "canonical transformation."

When we introduced the complex field variables we referred to the de Broglie-Schrödinger wave function. In this case, it is known that the wave functions ψ and $\psi \cdot e^{i\alpha}$ (α= real constant) describe the same physical situation. All observable quantities (as for instance $\psi^* \psi$) are independent of the phase constant α. We shall assume that our ψ-field is of the same kind. For this purpose we postulate that the differential Lagrange function (3.2) be invariant with respect to the substitution:

$$(3.10)\quad \psi \to \psi\, e^{i\alpha},\ \psi^* \to \psi^*\, e^{-i\alpha}\ (\alpha = \text{real constant})$$

The energy and momentum densities have, then, according to (3.8) the same invariance property.[1] We claim that such a ψ-field can be interpreted as a *charge-carrying* field, insofar as it is possible to define an electric charge density ρ and an electric current density s which satisfy a continuity equation. Indeed, if we let (ϵ = real constant):

$$(3.11)\quad \begin{cases} \varrho = -i\varepsilon\left(\frac{\partial L}{\partial \dot\psi}\psi - \frac{\partial L}{\partial \dot\psi^*}\psi^*\right) = -i\varepsilon(\pi\psi - \pi^*\psi^*), \\ s_k = -i\varepsilon\left(\frac{\partial L}{\partial \frac{\partial \psi}{\partial x_k}}\psi - \frac{\partial L}{\partial \frac{\partial \psi^*}{\partial x_k}}\psi^*\right). \end{cases}$$

then, using the classical field equations (3.3), it follows that:

$$\frac{\partial \varrho}{\partial t} + \operatorname{div} s = -i\varepsilon\left\{\left(\frac{\partial L}{\partial \psi}\psi + \sum_k \frac{\partial L}{\partial \frac{\partial \psi}{\partial x_k}}\frac{\partial \psi}{\partial x_k} + \frac{\partial L}{\partial \dot\psi}\dot\psi\right) - \left(\frac{\partial L}{\partial \psi^*}\psi^* + \sum_k \frac{\partial L}{\partial \frac{\partial \psi^*}{\partial x_k}}\frac{\partial \psi^*}{\partial x_k} + \frac{\partial L}{\partial \dot\psi^*}\dot\psi^*\right)\right\}.$$

[1] Pauli calls this invariance "gauge invariance of the first kind" (*Phys. Rev.* *58*, 716, 1940).

The right hand side of this equation is, up to a factor $(-\epsilon/\alpha)$, identical with the change of L (3.2) under the gauge transformation (3.10), for an infinitesimal change of phase α: $\psi \rightarrow \psi(1 + i\alpha)$, $\psi^* \rightarrow \psi^*(1 - i\alpha)$, i.e., $\delta\psi = i\alpha\psi$, $\delta\psi^* = -i\alpha\psi^*$. This variation of L vanishes because of the postulated invariance property, and hence the continuity equation for the electric densities follows:

$$\frac{\partial \varrho}{\partial t} + \nabla \cdot s = 0. \tag{3.12}$$

With $x_4 = i\,c\,t$, $s_4 = i\,c\,\rho$ one could also write instead of (3.11, 3.12):

$$s_\nu = -\,i\,\varepsilon\left(\frac{\partial L}{\partial \dfrac{\partial \psi}{\partial x_\nu}}\,\psi - \frac{\partial L}{\partial \dfrac{\partial \psi^*}{\partial x_\nu}}\,\psi^*\right), \quad \sum_\nu \frac{\partial s_\nu}{\partial x_\nu} = 0. \quad (\nu = 1\,..\,4) \tag{3.13}$$

In the quantized theory, ρ and s become Hermitian operators, and instead of (3.12) we obtain the operator equation:

$$\dot{\varrho}\,(x) \equiv \frac{i}{h}\,[H, \varrho\,(x)] = -\,\nabla \cdot s\,(x). \tag{3.14}$$

If there are several complex ψ-functions, $\psi, \psi' \ldots$, it may occur that L is only invariant with respect to a simultaneous gauge transformation:

$$\psi \rightarrow \psi\, e^{i\alpha}, \; \psi' \rightarrow \psi'\, e^{i\alpha}, \ldots$$

In this case the continuity equation (3.12) evidently holds only for the sum of the electric densities which correspond to the respective single fields (cf. §12).

Real field components ψ_σ, which can not be grouped together in pairs to complex fields, naturally have no transformation group of the kind (3.10). This is the reason that one can in general not construct density functions from such real field functions and their derivatives, which could be interpreted as electric charge and current density. (Applied to a real ψ-function the rule (3.11) results in: $\rho = 0$, $s_k = 0$.) Such real ψ-functions are used to describe electrically neutral fields like the electromagnetic field (light quanta are not charged). Charge-carrying fields, on the contrary, are represented appropriately by complex ψ-functions (cf. §§8 ff.).

§ 4. Lorentz Invariance. Time-Dependent Field Operators and Their Commutation Relations

In order to secure the Lorentz invariance of the classical formalism, it is only necessary to choose an invariant differential Lagrange function L. The field equations are then defined in an invariant way by the variational principle. For the integral $I = \int dt \int dx\mathbf{L}$, which is to be varied, is Lorentz-invariant on account of the invariance of the four-dimensional volume element $dt \cdot dx$ (the region of integration is an arbitrarily shaped space-time volume without preference of the time axis). The calculation of the Hamiltonian from the Lagrangian is effected formally alike in each coordinate system. Under a coordinate transformation $\mathbf{H} = -\mathbf{T}_{44}$ transforms like an energy density. Indeed, one verifies easily that $\mathbf{T}_{\mu\nu}$(2.8 or 3.8) transforms like a tensor of second rank, whereas s_ν (3.13) transforms like a four-vector.

The transformation properties of the field components ψ_σ determine the possible forms of the invariant function $\mathbf{L}$. It should be pointed out that (1.1) represents the most general expression for $\mathbf{L}$. Second and higher order space derivatives could appear only together with second and higher order time derivatives, which, however, would impair the canonical formalism.

The proof of the relativistic invariance is more difficult for the quantized theory. It is necessary first to remove the distinction of the time coordinate, which was introduced by considering space-dependent, but time-independent, operators $\psi_\sigma(x)$, $\pi_\sigma(x)$. We shall therefore choose more general operators $\psi_\sigma(x,t)$, $\pi_\sigma(x,t)$, which we construct again in analogy to particle mechanics.

In the quantum mechanics of systems of particles one would define the time-dependent operators $q_j(t)$, $p_j(t)$ by the requirement that they satisfy, as functions of time, the canonical equations of motion:

$$\frac{dq_j}{dt} = \frac{\partial H}{\partial p_j}, \qquad \frac{dp_j}{dt} = -\frac{\partial H}{\partial q_j}$$

and that for $t = 0$ they go over into the time-independent operators q_j, p_j. The commutation rules (1.7) must hold for every value of t since the time origin is arbitrary. One finds, indeed, by virtue of the equations of motion:

$$\frac{d}{dt}[p_j, q_{j'}] = -\left[\frac{\partial H}{\partial q_j}, q_{j'}\right] + \left[p_j, \frac{\partial H}{\partial p_{j'}}\right]$$

$$= -\frac{h}{i}\frac{\partial^2 H}{\partial p_{j'}\partial q_j} + \frac{h}{i}\frac{\partial^2 H}{\partial q_j \partial p_{j'}} = 0[1]$$

and similarly:

$$\frac{d}{dt}[q_j, q_{j'}] = \frac{d}{dt}[p_j, p_{j'}] = 0.$$

[1] It is known that:

$$[F, q_j] = \frac{h}{i}\frac{\partial F}{\partial p_j}, \quad [F, p_j] = -\frac{h}{i}\frac{\partial F}{\partial q_j}.$$

If the Hamilton operator does not contain the time explicitly ($\partial H/\partial t = 0$), one can construct the time-dependent operators from the time-independent ones with the help of the operators:

$$(4.1) \qquad S = e^{\frac{it}{h}H} = \sum_{k=0}^{\infty} \frac{1}{k!}\left(\frac{it}{h}H\right)^k, \qquad S^* = e^{-\frac{it}{h}H} = \sum_{k=0}^{\infty} \frac{1}{k!}\left(-\frac{it}{h}H\right)^k,$$

which are unitary because $SS^* = S^*S = 1$.[1] If one writes:

$$(4.2) \qquad q_j(t) = S\,q_j\,S^*, \qquad p_j(t) = S\,p_j\,S^*,$$

one has obviously $q_j(0) = q_j$, $p_j(0) = p_j$, and the time derivatives on account of:

$$\frac{dS}{dt} = S \cdot \frac{i}{h} H, \qquad \frac{dS^*}{dt} = -\frac{i}{h} H \cdot S^*$$

are:

$$(4.3) \qquad \begin{cases} \dfrac{dq_j(t)}{dt} = S \cdot \dfrac{i}{h}[H, q_j] \cdot S^* = S\dfrac{\partial H}{\partial p_j} S^* = \dfrac{\partial H}{\partial p_j}(t), \\[2ex] \dfrac{dp_j(t)}{dt} = S \cdot \dfrac{i}{h}[H, p_j] \cdot S^* = -S\dfrac{\partial H}{\partial q_j} S^* = -\dfrac{\partial H}{\partial q_j}(t), \end{cases}$$

The requirements which we have made for $q_j(t)$, $p_j(t)$ are thus satisfied. From the invariance of the commutation rules against the S-transformation (4.2) follows their validity for arbitrary values of time.

$$[q_j(t), q_{j'}(t)] = [p_j(t), p_{j'}(t)] = 0, \qquad [p_j(t), q_{j'}(t)] = \frac{h}{i}\delta_{jj'}.$$

In addition to this one can also calculate with the help of (4.1,.2) the commutators of quantities at different times (for instance, $[q_j(t), q_{j'}(t')]$).

In the field theory we shall analogously construct time-dependent operators $\psi_\sigma(x, t)$, $\pi_\sigma(x, t)$ from the operators $\psi_\sigma(x)$, $\pi_\sigma(x)$ which already depend on the space vector x as parameter. They shall satisfy the canonical field equations and shall go over, for $t = 0$, into the time-independent operators:

$$\psi_\sigma(x, 0) = \psi_\sigma(x), \qquad \pi_\sigma(x, 0) = \pi_\sigma(x).$$

In order to test the relativistic invariance of the commutation rules, one must proceed, generally speaking, in the following way: Under a Lorentz transformation $\overline{x_\mu} = \sum_\nu a_{\mu\nu}\, x_\nu$ the operators $\psi_\sigma(x, t)$, $\pi_\sigma(x, t)$ will have the same transformation properties as the corresponding classical field functions ψ_σ, π_σ. For instance if the classical field components ψ_ν form a 4-vector, the operators

[1] The time t must not be considered as an operator, but as a parameter (c-number). It commutes, therefore, with H.

$\bar{\psi}_\nu(\bar{x}, \bar{t})$ will be transformed cogrediently to the coordinates x_ν. The values of the commutators $[\psi_\sigma(x, t), \psi_{\sigma'}(x', t')]$ etc., determine the values of the commutators $[\bar{\psi}_\sigma(\bar{x}, \bar{t}), \bar{\psi}_{\sigma'}(\bar{x}', \bar{t}')$ etc., in the new frame of reference, so that the invariance can be tested directly. In particular one must require that the commutators of simultaneous quantities $(\bar{t} = \bar{t}')$ correspond in the new system of reference again to the canonical commutation rules (1.7).

Heisenberg and Pauli[1] have proved the invariance quite generally. In their proof they used the group property of the Lorentz transformations. On account of this property it is sufficient to prove the invariance under infinitesimal Lorentz transformations. This means a simplification for the following reason: For two events which are simultaneous in the new coordinate system, we have in the old system:

$$t - t' = \sum_{k=1}^{3} a_k (x_k - x_k'),$$

where now the a_k are supposed to be infinitesimal and need to be considered only in the first order. In this approximation it is sufficient to calculate, for instance:

$$[\psi_\sigma(x, t), \psi_{\sigma'}(x', t')] = \left[\left\{\psi_\sigma(x, t') + (t - t')\left(\frac{\partial \psi_\sigma(x, t)}{\partial t}\right)_{t=t'}\right\}, \psi_{\sigma'}(x', t')\right]$$
$$= (t - t') \cdot [\dot{\psi}_\sigma(x), \psi_{\sigma'}(x')]$$

which can be done with the help of the canonical field equations and the commutation rules (1.7). We forego here a general proof for the invariance. We shall rather derive a few more general formulas for the commutators of time-dependent field operators. These formulas will enable us to give a Lorentz invariant formulation to the commutation rules for certain special types of fields, which we are going to consider in the following chapters.

We define:

$$\psi_\sigma(x, t) = e^{\frac{it}{h}E} \psi_\sigma(x) e^{-\frac{it}{h}E}, \quad \pi_\sigma(x, t) = e^{\frac{it}{h}E} \pi_\sigma(x) e^{-\frac{it}{h}E}. \tag{4.4}$$

The expression (4.4) is similar to the expression (4.1,.2) if one identifies E with the integral Hamilton function $H = \int dx \mathsf{H}$. But we make the reservation that we can choose for E an energy operator which is different from H, for instance,

[1] *Z. Phys.* *56*, 1, 1929. Cf. also Rosenfeld, *ibid.* *63*, 574, 1930.

the Hamiltonian of an "unperturbed system."[1] The assumption is essential, however, that the operator E does not contain the time explicitly:

$$\frac{\partial E}{\partial t} = 0. \tag{4.5}$$

By differentiating with respect to t (for constant x, indicated by writing $\partial/\partial t$), it follows from (4.4), analogously to (4.3):

$$\begin{cases} \dfrac{\partial \psi_\sigma(x,t)}{\partial t} = e^{\frac{it}{h}E} \cdot \dfrac{i}{h}[E, \psi_\sigma(x)] \cdot e^{-\frac{it}{h}E} = \dfrac{i}{h}[E, \psi_\sigma(x,t)], \\ \dfrac{\partial \pi_\sigma(x,t)}{\partial t} = e^{\frac{it}{h}E} \cdot \dfrac{i}{h}[E, \pi_\sigma(x)] \cdot e^{-\frac{it}{h}E} = \dfrac{i}{h}[E, \pi_\sigma(x,t)]. \end{cases} \tag{4.6}$$

We consider now the commutators $[\psi_\sigma(x,t), \psi_{\sigma'}(x',t')]$. Using the identity:

$$\psi_\sigma(x,t) = e^{\frac{it'}{h}E}\, \psi_\sigma(x, t-t')\, e^{-\frac{it'}{h}E}$$

we can write:

$$[\psi_\sigma(x,t), \psi_{\sigma'}(x',t')] = e^{\frac{it'}{h}E}\, [\psi_\sigma(x, t-t'), \psi_{\sigma'}(x')]\, e^{-\frac{it'}{h}E}.$$

If we expand here $\psi_\sigma(x, t-t')$ in powers of $t-t'$:

$$\psi_\sigma(x, t-t') = \sum_{n=0}^{\infty} \frac{1}{n!}(t-t')^n \cdot \overset{(n)}{\psi_\sigma}(x),$$

where:

$$\overset{(n)}{\psi_\sigma}(x) \equiv \left(\frac{\partial^n \psi_\sigma(x,t)}{\partial t^n}\right)_{t=0}, \tag{4.7}$$[2]

it follows:

$$[\psi_\sigma(x,t), \psi_{\sigma'}(x',t')] = e^{\frac{it'}{h}E} \sum_{n=0}^{\infty} \frac{1}{n!}(t-t')^n \left[\overset{(n)}{\psi_\sigma}(x), \psi_{\sigma'}(x')\right] e^{-\frac{it'}{h}E}. \tag{4.8}$$

Here the assumption shall be made, applicable to all later examples, that the

[1] This is of importance for the "multiple-time field theories"; cf. §18 (quantum electrodynamics, with multiple-time variables).

[2] In the case $E = H$, according to (4.6) it is true that:

$$\overset{(1)}{\psi} = \dot{\psi}, \quad \overset{(2)}{\psi} = \ddot{\psi}, \quad \ldots.$$

energy density **E** (where $E = \int dx\ \mathbf{E}$) is a quadratic form of the $\psi_\sigma(x)$, $\pi_\sigma(x)$ and their space derivatives. In this case, it is easy to see that the commutators $[E, \psi_\sigma(x)]$ and $[E, \pi_\sigma(x)]$ are, on account of the commutation rules (1.7), homogeneous linear functions of the $\psi_\sigma(x)$, $\pi_\sigma(x)$ and their space derivatives. Consequently, the operators $\partial\psi_\sigma(x, t)/\partial t$ and $\partial\pi_\sigma(x, t)/\partial t$ as well as the higher derivatives $\partial^n\psi_\sigma(x, t)/\partial t^n$, $\partial^n\pi_\sigma(x, t)/\partial t^n$ are, according to (4.6 and .4), also linear homogeneous functions of the $\psi_\sigma(x, t)$, $\pi_\sigma(x, t)$ and their space derivatives. Hence we can write:

$$\frac{\partial^n\psi_\sigma(x, t)}{\partial t^n} = \sum_{\sigma'} \{c^{(n)}_{\sigma\sigma'} \cdot \psi_{\sigma'}(x, t) + d^{(n)}_{\sigma\sigma'} \cdot \pi_{\sigma'}(x, t)\}, \tag{4.9}$$

where $c^{(n)}_{\sigma\sigma'}$, $d^{(n)}_{\sigma\sigma'}$ are certain differential operators with respect to the space coordinates x. In particular we have at $t = 0$, for the operators $\overset{(n)}{\psi}_\sigma(x)$, defined by (4.7):

$$\overset{(n)}{\psi}_\sigma(x) = \sum_{\sigma'} \{c^{(n)}_{\sigma\sigma'} \cdot \psi_{\sigma'}(x) + d^{(n)}_{\sigma\sigma'} \cdot \pi_{\sigma'}(x)\}. \tag{4.10}$$

From this we obtain for the commutators $\left[\overset{(n)}{\psi}_\sigma(x), \psi_{\sigma'}(x')\right]$ which occur in (4.8) on account of the commutation rules (1.7):

$$\left[\overset{(n)}{\psi}_\sigma(x), \psi_{\sigma'}(x')\right] = \frac{h}{i}\, d^{(n)}_{\sigma\sigma'} \cdot \delta(x - x'); \tag{4.11}$$

The differential operator $d^{(n)}_{\sigma\sigma'}$ acts here upon the space coordinates x in the argument of the δ-function. [The singular δ-function is here considered to be approximated by a regular function of $(x - x')$. In a space integral $\int dx\, f(x) \cdot d^{(n)}_{\sigma\sigma'}\, \delta(x - x')$ the differentiations can be carried out by partial integration, and after this the limiting process to the singular δ-function can be carried out according to (1.9)]. It is essential in equation (4.11) that its right hand side is independent of the field variables, i.e., that it commutes with E. Introducing (4.11) into (4.8), we can place the factors $e^{\pm\frac{it'}{h}E}$ therefore either to the right or to the left so that they cancel each other. The result is:

$$\begin{aligned}[\psi_\sigma(x, t), \psi_{\sigma'}(x', t')] &= \frac{h}{i} \sum_{n=0}^{\infty} \frac{1}{n!} (t - t')^n\, d^{(n)}_{\sigma\sigma'}\, \delta(x - x') \\ &\equiv \frac{h}{i} D_{\sigma\sigma'}(x - x', t - t').\end{aligned} \tag{4.12}$$

The commutators of the time-dependent operators ψ_σ are thus (singular) functions of $x - x'$ and $t - t'$, which can be computed as soon as the differential operators $d^{(n)}_{\sigma\sigma'}$, defined by (4.9 or 4.10), are known.

In the same way one could also calculate the commutators $[\pi_\sigma(x, t), \pi_{\sigma'}(x', t')]$ and $[\pi_\sigma(x, t), \psi_{\sigma'}(x', t')]$. This is, however, not necessary since all statements of the canonical commutation rules are already contained completely in (4.12). For examining the Lorentz invariance, the consideration of (4.12) alone is sufficient. In order to show this we differentiate in (4.12) both sides n times with respect to t and n' times with respect to t' and we set: $t = t' = 0$. One obtains according to (4.7):

$$\left[\overset{(n)}{\psi_\sigma}(x), \overset{(n')}{\psi_{\sigma'}}(x')\right] = \frac{h}{i}(-1)^{n'} d^{(n+n')}_{\sigma\sigma'} \delta(x - x'). \tag{4.13}$$

With $n = n' = 0 \left(\overset{(0)}{\psi_\sigma} = \psi_\sigma\right)$, it follows simply:

$$[\psi_\sigma(x), \psi_{\sigma'}(x')] = \frac{h}{i} d^{(0)}_{\sigma\sigma'} \delta(x - x') = 0, \tag{4.14}$$

since according to (4.9):

$$d^{(0)}_{\sigma\sigma'} = 0 \tag{4.15}$$

(and $c^{(0)}_{\sigma\sigma'} = \delta_{\sigma\sigma'}$). If one sets in (4.13) $n = 1, n' = 0$ or $n = n' = 1$, and if one expresses the $\overset{(1)}{\psi_\sigma}$ according to (4.10) by the ψ_σ and π_σ, one obtains linear equations for the commutators $[\pi_\sigma(x), \psi_{\sigma'}(x')]$ and $[\pi_\sigma(x), \pi_{\sigma'}(x')]$, which in connection with (4.14) are just sufficient for their determination. In this way one can recover the canonical commutation rules (1.7); i.e., there exists a complete equivalence between these and the new relations (4.12).

In order to prove now the relativistic invariance of the canonical quantization method, one has only to verify that the relations (4.12) hold in all Lorentz systems, which means that the functions $D_{\sigma\sigma'}(x - x', t - t')$ in (4.12) transform like the products $\psi_\sigma(x, t) \cdot \psi_{\sigma'}(x', t')$ [for instance, like the components of a tensor of the second rank, in case the ψ_σ $(\sigma = 1 \ldots 4)$ form a 4-vector]. In the following we shall carry out the calculation of the functions $D_{\sigma\sigma'}$ for various special types of fields; and in doing so the invariance will be apparent each time in the construction of the formulas.

We can carry the calculation of the commutators (4.12) a step further, using the fact that the Hamiltonian E in all the problems which are of interest is such that the ψ_σ as space-time functions satisfy the Schrödinger-Gordon wave equation:[1]

$$(4.16) \qquad (\Box - \mu^2)\,\psi_\sigma(x,t) = 0, \quad \text{wo } \Box \equiv \sum_\nu \frac{\partial^2}{\partial x_\nu^2} \equiv \nabla^2 - \frac{1}{c^2}\frac{\partial^2}{\partial t^2}.$$

It is known that for instance the de Broglie wave functions:

$$e^{i(kx \mp \sqrt{\mu^2 + k^2}\,ct)}$$

are (complex) particular solutions of this wave equation. The scalar constant $\mu\, h/c$ represents the rest mass of the respective particles. It will turn out, indeed, in the following that the stationary states of quantized fields obeying the field equations (4.16) represent systems of particles with the rest mass $\mu\, h/c$.

Let us now consider (4.16) to hold for the operators $\psi_\sigma(x,t)$:

$$(4.17) \qquad \frac{\partial^2 \psi_\sigma(x,t)}{\partial t^2} = c^2 (\nabla^2 - \mu^2)\, \psi_\sigma(x,t).$$

Differentiating the equation (4.9) twice with respect to t, it follows:

$$\frac{\partial^{n+2}\psi_\sigma(x,t)}{\partial t^{n+2}} \equiv \sum_{\sigma'} \{c_{\sigma\sigma'}^{(n+2)}\psi_{\sigma'}(x,t) + d_{\sigma\sigma'}^{(n+2)}\pi_{\sigma'}(x,t)\}$$

$$= c^2(\Delta - \mu^2) \sum_{\sigma'} \{c_{\sigma\sigma'}^{(n)}\psi_{\sigma'}(x,t) + d_{\sigma\sigma'}^{(n)}\pi_{\sigma'}(x,t)\},$$

i.e.:

$$(4.18) \qquad d_{\sigma\sigma'}^{(n+2)} = d_{\sigma\sigma'}^{(n)}\, c^2(\nabla^2 - \mu^2)$$

(similarly for the operators $c_{\sigma\sigma'}^{(n)}$, which are of no interest here). Since $d_{\sigma\sigma'}^{(0)}$ is zero by definition, all $d_{\sigma\sigma'}^{(n)}$ with even n vanish according to (4.18):

$$(4.19) \qquad d_{\sigma\sigma'}^{(2m)} = 0$$

While for odd n:

$$(4.20) \qquad d_{\sigma\sigma'}^{(2m+1)} = d_{\sigma\sigma'}^{(1)}\, [c^2(\nabla^2 - \mu^2)]^m.$$

Here the operator $d_{\sigma\sigma'}^{(1)}$ can be defined according to (4.9 or 4.10) and (4.6) by the equation:

$$(4.21) \qquad \overset{(1)}{\psi_\sigma}(x) \equiv \frac{i}{h}[E, \psi_\sigma(x)] = \sum_{\sigma'} \{c_{\sigma\sigma'}^{(1)}\psi_{\sigma'}(x) + d_{\sigma\sigma'}^{(1)}\pi_{\sigma'}(x)\}$$

[1] We are here dealing only with the wave equation in the "field free case." Cf. §11 for the general Schrödinger-Gordon equation.

With the help of (4.19,.20) one obtains for the functions $D_{\sigma\sigma'}$, defined by (4.12):

$$D_{\sigma\sigma'}(x, t) \equiv \sum_{n=0}^{\infty} \frac{1}{n!} t^n d^{(n)}_{\sigma\sigma'} \delta(x) = d^{(1)}_{\sigma\sigma'} D(x, t), \tag{4.22}$$

where:

$$D(x, t) = \sum_{m=0}^{\infty} \frac{1}{(2m+1)!} t^{2m+1} [c^2 (\nabla^2 - \mu^2)]^m \delta(x). \tag{4.23}$$

In order to carry out the summation with respect to *m*, we write Dirac's δ-function as a Fourier integral:

$$\delta(x) = \frac{1}{(2\pi)^3} \int dk\, e^{ikx} \tag{4.24}$$

(kx = scalar product of the wave number vector k with the space vector x; dk = volume element in k-space). This integral representation of the δ-function is a consequence of the Fourier integral theorem.

$$\int dx\, f(x)\, \delta(x - x') = \frac{1}{(2\pi)^3} \int dk \int dx\, f(x)\, e^{ik(x-x')} = f(x').$$

This equation agrees with (1.9) and is equivalent to (1.8). Strictly speaking, one should substitute into the integrand in (4.24) a convergence factor (for instance, const. $e^{-\alpha k^2}$), which corresponds to replacing the singular δ-function by a regular function, and then carry out the limiting process ($\alpha \to 0$) only in the space integral $\int dx f(x)\, \delta(x - x')$. Such factors, which suppress ("cut off") the contributions of the large values of $|k|$, should always be included in the following wherever necessary.

From (4.23) we get with the help of (4.24):

$$D(x, t) = \frac{1}{(2\pi)^3} \int dk\, e^{ikx} \sum_{m=0}^{\infty} \frac{1}{(2m+1)!} t^{2m+1} [-c^2 (k^2 + \mu^2)]^m$$

or, since:

$$\sum_{m=0}^{\infty} \frac{1}{(2m+1)!} (-1)^m \tau^{2m+1} = \sin \tau:$$

$$D(x, t) = \frac{1}{(2\pi)^3} \int dk\, e^{ikx} \frac{\sin\left(tc\sqrt{\mu^2 + k^2}\right)}{c\sqrt{\mu^2 + k^2}}. \tag{4.25}$$

We claim that the function $D(x - x', t - t')$ is invariant under Lorentz transformations. For the proof, we consider the space-time Fourier integral:

$$\int dk \int dk_0\, g\,(k_0^2 - k^2)\; e^{i(kx - k_0 ct)}, \tag{4.26}$$

where again we have written x, instead of $x - x'$, and t, instead of $t - t'$. The 3-vector k together with k_0 forms a 4-vector. The wave phase $(kx - k_0\, ct)$, as well as the four-dimensional volume element $dk \cdot dk_0$ are consequently Lorentz-invariant. For the same reason the Fourier integral (4.26) is invariant, provided the amplitude function g depends only on the invariant $k_0^2 - k^2$. This is even true if the integral is extended only over an invariant subspace of the k,k_0-space, as, for instance, over the cone $k_0 > |k|$, or over the anticone $k_0 < -|k|$.[1] For these two cases we shall rewrite the integral (4.26) by introducing as integration variable $\mu = \sqrt{k_0^2 - k^2}$ instead of k_0 (for constant vector k): $k_0 = \pm\sqrt{\mu^2 + k^2}$, $dk_0 = d\mu\ \ \mu/\sqrt{\mu^2 + k^2}$. Interchanging the order of integrations with respect to μ and the k-space, one obtains for (4.26):

$$\int_0^\infty d\mu\, \mu\, g\,(\mu^2) \int dk \frac{1}{\sqrt{\mu^2 + k^2}}\; e^{i(kx \mp \sqrt{\mu^2 + k^2}\, ct)}.$$

Since the integration variable μ $(= \sqrt{k_0^2 - k^2})$ is Lorentz-invariant and g is an arbitrary function (one could choose in its place a one-dimensional δ-function, which is different from zero for one particular value only), it follows from the invariance of the above integrals, that the two space-time functions:

$$\int dk \frac{1}{\sqrt{\mu^2 + k^2}}\; e^{i(kx \mp \sqrt{\mu^2 + k^2}\, ct)} \tag{4.27}$$

are invariant for every (real) value of the parameter μ. The difference of these two functions is, apart from a numerical factor, the function $D\,(x,\, t)$ (4.25). With this the Lorentz invariance of this function is established.[2]

If the wave equation (4.16) holds, the commutators (4.12) according to (4.22) may be written in the form:

$$[\psi_\sigma\,(x,\, t),\, \psi_{\sigma'}\,(x',\, t')] = \frac{h}{i}\, d^{(1)}_{\sigma\sigma'}\, D\,(x - x',\, t - t') \tag{4.28}$$

---

[1] Transformations reversing the time axis are not considered here.

[2] The invariant D-function with the special value $\mu = 0$ was first introduced by Jordan and Pauli (*Z. Phys.* 47, 151, 1928) and was used to formulate the invariant commutation rules for the electromagnetic field strengths. Cf. §§16 and 18.

In order to establish their relativistic invariance it is sufficient to show that the operators $d^{(1)}_{\sigma\sigma'}$, defined by (4.21), transform the same way as the (classical) functions $\psi_\sigma \cdot \psi_{\sigma'}$. This can easily be done in the following for each particular field which we shall consider.

It should be pointed out that all the above mentioned formulas can also be applied to complex field functions or to non-Hermitian operators ψ_σ. If one treats therefore the ψ and the ψ^* as independent field variables, as in §3, one can identify in (4.28, 4.21) one half of the ψ_σ with ψ, and the other half with ψ^* (cf. §§8 and 12).

The invariant function $D(x, t)$ can be written as a Bessel-function of the argument $\mu \sqrt{c^2 t^2 - x^2}$ by carrying out the k-space integration in (4.25). In addition to that, there exists a singularity of the δ-type on the light cone $c^2 t^2 - x^2 = 0$.[1] We shall not dwell further on this point, but restrict ourselves to stating some characteristic properties of the D-function, which we shall use later on. $D(x, t)$ is, according to (4.25), a real function with the symmetry properties:

$$D(x, t) = D(-x, t) = -D(x, -t) = -D(-x, -t). \tag{4.29}$$

Since the Fourier integral (4.25) represents a superposition of de Broglie waves, $e^{i(kx \mp \sqrt{\mu^2 + k^2}\, ct)}$, which satisfy the Schrödinger-Gordon wave equation (4.16), the D-function is also a solution of this differential equation:

$$(\Box - \mu^2)\, D(x, t) = 0. \tag{4.30}$$

For $t = 0$, (4.25 or 4.29) yields:

$$D(x, 0) = 0. \tag{4.31}$$

Since D is an invariant, this means that D vanishes outside the light cone $x^2 - c^2 t^2 = 0$, for each world point of this domain can be changed to a point $t = 0$ by a Lorentz transformation:

$$D(x, t) = 0 \text{ für } c\,|t| < |x|. \tag{4.32}$$

Differentiating in (4.23 or 4.25) with respect to t and putting $t = 0$, one finds the further important equation [cf. (4.24)]:

$$\left(\frac{\partial D(x, t)}{\partial t}\right)_{t=0} = \delta(x). \tag{4.33}$$

[1] Cf. Dirac, *Proc. Cambridge Phil. Soc.* 30, 150, 1934; Pauli, *Phys. Rev.* 58, 716, 1940. Regarding the special case $\mu = 0$, cf. §18.

It is well known that in quantum theory there exists a connection between the commutators of two operators and the limits of accuracy for the measurements of the two corresponding physical quantities. The accuracy of measurement is unlimited only if both operators commute, i.e., if they can be simultaneously transformed into diagonal form. Thus the commutation rules (4.12 and 4.28) imply that the knowledge of the function ψ_σ at the world point x t and of $\psi_{\sigma'}$ at the world point x', t' is limited according to the uncertainty relation:

$$\delta\psi_\sigma (x, t) \;\; \delta\psi_{\sigma'} (x', t') \gtrsim h \, D_{\sigma\sigma'} (x - x', t - t').$$

The D-function has on the light cone $|x - x'|^2 = c^2 (t - t')^2$ a δ singularity. On account of the fact that only integrals over such singularities have a direct meaning, it is more consistent to consider averages of the field values over the space-time regions Γ, Γ', instead of their values at certain space-time points. Indicating this averaging operation by M_Γ, $M_{\Gamma'}$, respectively, we may write the uncertainty relations rigorously:

$$\delta \left\{M_\Gamma \psi_\sigma (x, t)\right\} \;\; \delta \left\{M_{\Gamma'} \psi_{\sigma'} (x', t')\right\} \gtrsim h \;\, M_\Gamma M_{\Gamma'} D_{\sigma\sigma'} (x - x', \; t - t'). \tag{4.34}$$

For the case of the electromagnetic field strength Bohr and Rosenfeld[1] have discussed hypothetical experiments which would allow a measurement of the field with the optimum accuracy that could be expected according to (4.34). The experimental errors, which principally cannot be eliminated, arise from the field radiated by the electrically charged test bodies during a measurement of their momentum change, which has to be made in order to determine the field strength. This field cannot be compensated or determined with unlimited accuracy. In order to reach the greatest accuracy it is essential to use as test bodies not point charges but extended rigid bodies, the atomic structure of which can be ignored.

In case that all point pairs of the domain Γ, Γ' are connected by spacelike vectors ($|x - x'| > c|t - t'|$), we have, according to (4.28 and 4.32):

$$\delta \left\{M_\Gamma \psi_\sigma (x, t)\right\} \;\; \delta \left\{M_{\Gamma'} \psi_{\sigma'} (x', t')\right\} = 0;$$

In this case the two field measurements do not disturb each other. This is to be expected, since the field disturbances, caused by a measurement, can at most propagate with the velocity of light and hence could not be observed in the other world domain.

---

[1] *Kgl. Danske Vidensk. Selsk., Math.-fys. Medd. XII*, 8, 1933. Cf. also Heitler, *Quantum Theory of Radiation*, Oxford Univ. Press, London, 1936, §8.

§ 5. Introduction of the Momentum Space

For later calculations we shall often need spatial Fourier decompositions of the field functions ψ_σ, or of the corresponding operators. It will be convenient to use Fourier series instead of the more general Fourier integrals, so that the wave-number vectors form a countable set. For that purpose we restrict the space functions by a periodicity condition. The domain of periodicity shall be a cube with edges of length l, i.e., with a volume $V = l^3$. The plane waves e^{ikx} (kx again means the scalar product of the 3-vectors k and x) with this periodicity are characterized by wave-number vectors k, the cartesian components of which are integral multiples of $2\pi/l$. Hence the periodicity condition changes the continuous k-space into a cubical point lattice with the lattice constant $2\pi/l$ and the cell volume $(2\pi/l)^3$. We shall write the Fourier expansion of the field functions ψ_σ and π_σ as follows:

$$(5.1)\qquad \psi_\sigma(x) = V^{-1/2}\sum_k q_{\sigma,k}\, e^{ikx}, \qquad \pi_\sigma(x) = V^{-1/2}\sum_k p_{\sigma,k}\, e^{-ikx};$$

Here the summations are to be extended over all lattice points of the k-space. The Fourier coefficients $q_{\sigma,k}$, $p_{\sigma,k}$ are in the classical theory functions of time. In quantum theory they are operators which do not contain the time explicitly if the same is true of the operators $\psi_\sigma(x)$, $\pi_\sigma(x)$. For real (Hermitian) ψ_σ, π_σ, one must demand that:

$$(5.2)\qquad q_{\sigma,-k} = q^*_{\sigma,k}, \quad p_{\sigma,-k} = p^*_{\sigma,k}$$

that is, $q_{\sigma,-k}$, $p_{\sigma,-k}$ are complex conjugate (Hermitian conjugate) to the $q_{\sigma,k}$, $p_{\sigma,k}$. This is necessary and sufficient for:

$$\psi^*_\sigma(x) = V^{-1/2}\sum_k q^*_{\sigma,k}\, e^{-ikx} = \psi_\sigma(x),$$

$$\pi^*_\sigma(x) = V^{-1/2}\sum_k p^*_{\sigma,k}\, e^{+ikx} = \pi_\sigma(x).$$

One obtains the inverse formulas to (5.1) by multiplication with $V^{-1/2}\, e^{\mp ik'x}\, dx$ and integration over the periodicity cube V:

$$(5.3)\qquad q_{\sigma,k} = V^{-1/2}\int_V dx\, \psi_\sigma(x)\, e^{-ikx}, \qquad p_{\sigma,k} = V^{-1/2}\int_V dx\, \pi_\sigma(x)\, e^{ikx}.$$

The canonical commutation rules (1.7) can be retained by restricting them to point pairs x, x', which lie in the same periodicity cube V. On account of the periodicity they are then determined also for other point pairs. Their application to (5.3) shows that the $q_{\sigma,k}$ and the $p_{\sigma,k}$ commute among themselves. Further:

$$[p_{\sigma,k}, q_{\sigma' k'}] = V^{-1} \int_V dx \int_V dx' \, [\pi_\sigma(x), \psi_{\sigma'}(x')] \, e^{i(kx - k'x')}$$

$$= \frac{h}{i} \delta_{\sigma\sigma'} V^{-1} \int_V dx \, e^{i(k-k')x} = \frac{h}{i} \delta_{\sigma\sigma'} \delta_{kk'}.$$

Thus the commutation rules of the q_j, p_j, correspond to those of canonically conjugate variables:

$$[q_{\sigma,k}, q_{\sigma',k'}] = [p_{\sigma,k}, p_{\sigma',k'}] = 0, \quad [p_{\sigma,k}, q_{\sigma',k'}] = \frac{h}{i} \delta_{\sigma\sigma'} \delta_{kk'}; \tag{5.4}$$

the equations (5.1) furnish therefore a canonical transformation from the old variables $\psi_\sigma(x)$, $\delta x \cdot \pi_\sigma(x)$ (cf. §1) to the new variables $q_{\sigma,k}$, $p_{\sigma,k}$.

The Fourier decomposition of complex, non-Hermitian field functions ψ (cf. §3) could, of course, be carried out in the same way according to the equations (5.1,.3). The only difference is that the condition of reality or of the Hermitian character (5.2) is no longer needed. Referring to the notations of §3, we write:

$$\left\{ \begin{aligned} \psi(x) &= V^{-1/2} \sum_k q_k e^{ikx}, & \pi(x) &= V^{-1/2} \sum_k p_k e^{-ikx}, \\ \psi^*(x) &= V^{-1/2} \sum_k q_k^* e^{-ikx}, & \pi^*(x) &= V^{-1/2} \sum_k p_k^* e^{ikx}. \end{aligned} \right. \tag{5.5}$$

From the commutation rules (3.9), it follows, as before:

$$[p_k, q_{k'}] = [p_k^*, q_{k'}^*] = \frac{h}{i} \delta_{kk'}, \tag{5.6}$$

while all other pairs of variables commute.

The restriction to periodic field functions can easily be removed by carrying out the limiting process $l \to \infty$. In this way the k-lattice changes into a continuum, and the Fourier series become Fourier integrals. In the commutation rules the δ-function $\delta(k - k')$ in k-space appears instead of $\delta_{kk'}$. We shall not enter into this, however.

Chapter II

Scalar Fields

§ 6. Real Field in Vacuum

We shall first discuss, as the simplest example, a real scalar field, i.e., a real Lorentz-invariant field with one component $\psi_1 = \psi$. The Schrödinger-Gordon wave equation (4.16) shall be considered as the classical field equation.

$$(\Box - \mu^2)\,\psi \equiv -\frac{1}{c^2}\,\ddot{\psi} + \nabla^2\,\psi - \mu^2\,\psi = 0. \tag{6.1}$$

This is achieved by choosing the following, evidently Lorentz-invariant Lagrangian:

$$L = -\frac{1}{2}\,c^2\left\{\sum_\nu\left(\frac{\partial\psi}{\partial x_\nu}\right)^2 + \mu^2\,\psi^2\right\} = \frac{1}{2}\left\{\dot{\psi}^2 - c^2\,|\,\nabla\,\psi\,|^2 - c^2\,\mu^2\,\psi^2\right\}. \tag{6.2}$$

Indeed, we have:

$$\frac{\partial L}{\partial\dot{\psi}} = \dot{\psi}, \qquad \frac{\partial L}{\partial\dfrac{\partial\psi}{\partial x_k}} = -\,c^2\,\frac{\partial\psi}{\partial x_k}, \qquad \frac{\partial L}{\partial\psi} = -\,c^2\,\mu^2\,\psi,$$

so that the field equation (1.2) takes the form (6.1). The field π canonically conjugate to ψ is according to (1.4):

$$\pi = \frac{\partial L}{\partial\dot{\psi}} = \dot{\psi}, \tag{6.3}$$

and it follows for the differential Hamiltonian from (1.5):

$$H = \frac{1}{2}\left\{\pi^2 + c^2\,|\,\nabla\,\psi\,|^2 + c^2\,\mu^2\,\psi^2\right\}. \tag{6.4}$$

One may notice that H is positive-definite, as should be required for an energy density.

In the quantized theory ψ and π become space-dependent, Hermitian operators ($\psi^* = \psi$, $\pi^* = \pi$) with the commutators [cf. (1.7, 1.8)]:

$$(6.5)\quad [\psi(x), \psi(x')] = [\pi(x), \pi(x')] = 0, \qquad [\pi(x), \psi(x')] = \frac{h}{i}\,\delta(x - x').$$

Substituting these operators into $\boldsymbol{H}$ (6.4) and $H = \int dx\, \boldsymbol{H}$ we get the Hermitian Hamiltonian operator. We shall formulate the canonical field equations according to (1.12) and for this purpose we calculate first the commutators of $\boldsymbol{H}$ (6.4) with ψ and π:

$$[\boldsymbol{H}(x), \psi(x')] = \frac{1}{2}\,[(\pi(x))^2, \psi(x')]$$

$$= \frac{1}{2}\,\{\pi(x)\cdot[\pi(x), \psi(x')] + [\pi(x), \psi(x')]\cdot\pi(x)\}$$

$$= \pi(x)\cdot\frac{h}{i}\,\delta(x - x'),$$

$$[\boldsymbol{H}(x), \pi(x')] = \frac{c^2}{2}\Big[\{|\nabla\psi(x)|^2 + \mu^2(\psi(x))^2\}, \pi(x')\Big]$$

$$= -c^2\frac{h}{i}\{\nabla\psi(x)\cdot\nabla\delta(x - x') + \mu^2\psi(x)\,\delta(x - x')\}.$$

From this we obtain by integration over the x-space $[H, \psi(x')]$ and $[H, \pi(x')]$. By partial integration:

$$\int dx\,\nabla\psi(x)\cdot\nabla\delta(x - x') = -\int dx\,\nabla^2\psi(x)\,\delta(x - x')$$

and with the help of (1.9) it follows:

$$(6.6)\quad \begin{cases} \dot{\psi}(x) \equiv \dfrac{i}{h}\,[H, \psi(x)] = \pi(x) \quad [\text{vgl. } (6.3)], \\ \dot{\pi}(x) \equiv \dfrac{i}{h}\,[H, \pi(x)] = c^2\{\nabla^2\psi(x) - \mu^2\psi(x)\}. \end{cases}$$

We can eliminate π by forming $\ddot{\psi} \equiv i/h\;[H, \dot{\psi}]$:

$$(6.7)\quad \ddot{\psi}(x) = \dot{\pi}(x) = c^2\{\nabla^2\psi(x) - \mu^2\psi(x)\}.$$

This operator equation corresponds formally to the classical field equation (6.1). It states, in particular, that the expectation value of ψ as space-time function obeys the wave equation (6.1) (cf. footnote 1, p. 7).

In order to prove in this case the relativistic invariance of the canonical quantization method, we construct the time-dependent operator $\psi(x,t)$ according to (4.4), using here $E = H = \int dx\, \mathsf{H}$:

$$\psi(x, t) = e^{\frac{it}{h}H}\, \psi(x)\, e^{-\frac{it}{h}H}.$$

According to (4.6) and (6.7):

$$\frac{\partial^2 \psi(x, t)}{\partial t^2} = e^{\frac{it}{h}H}\, \ddot{\psi}(x)\, e^{-\frac{it}{h}H} = c^2 (\nabla^2 - \mu^2)\, \psi(x, t),$$

i.e., $\psi(x, t)$ satisfies the Schrödinger-Gordon wave equation. It was proved in §4, that in this case the canonical commutation rules (6.5) are equivalent to (4.28), i.e.:

$$[\psi(x, t), \psi(x', t')] = \frac{h}{i}\, d^{(1)}\, D(x - x', t - t'),$$

where $d^{(1)}$ is determined by (4.21):

$$\overset{(1)}{\psi}(x) \equiv \dot{\psi}(x) = c^{(1)} \cdot \psi(x) + d^{(1)} \cdot \pi(x).$$

A comparison with (6.6) yields: $d^{(1)} = 1$ (and $c^{(1)} = 0$), hence:

$$[\psi(x, t), \psi(x', t')] = \frac{h}{i}\, D(x - x', t - t'). \tag{6.8}$$

Here both sides of the equation are invariant under Lorentz transformations, which is what we set out to prove.

We shall verify in addition for this simple case that the canonical commutation rules follow from the invariant ones (6.8). Since (6.8) is equivalent to (4.12), the equations (4.13) derived from it are also valid:

$$\left[\overset{(n)}{\psi}(x), \overset{(n')}{\psi}(x')\right] = \frac{h}{i}\, (-1)^{n'}\, d^{(n+n')}\, \delta(x - x'),$$

where the $d^{(n+n')}$ are determined by (4.19, 4.20), with $d^{(1)} = 1$. We have, since $d^{(2m)} = 0$, for $n = n'$:

$$\left[\overset{(n)}{\psi}(x), \overset{(n)}{\psi}(x')\right] = 0,$$

hence, on account of $\overset{(0)}{\psi} = \psi$, $\overset{(1)}{\psi} = \dot{\psi} = \pi$:

$$[\psi(x), \psi(x')] = 0, \quad [\pi(x), \pi(x')] = 0.$$

One obtains on the other hand, with $n = 1$, $n' = 0$:

$$[\pi(x), \psi(x')] = \frac{h}{i}\,\delta(x - x'),$$

With this all the canonical commutation rules (6.5) are obtained again. In the following it will no longer be necessary to use the time-dependent operator $\psi(x, t)$.

The physical statements of the quantized theory are obtained from the energy-momentum tensor. For the canonical tensor (2.8) one obtains on account of the Lagrangian (6.2):

$$T_{\mu\nu} = c^2 \frac{\partial\psi}{\partial x_\mu}\frac{\partial\psi}{\partial x_\nu} + L\,\delta_{\mu\nu}.$$

The following order of the factors will make it Hermitian:

$$T_{\mu\nu} = \frac{c^2}{2}\left(\frac{\partial\psi}{\partial x_\mu}\frac{\partial\psi}{\partial x_\nu} + \frac{\partial\psi}{\partial x_\nu}\frac{\partial\psi}{\partial x_\mu}\right) + L\,\delta_{\mu\nu}; \tag{6.9}$$

For instance the momentum density operator becomes:

$$G_k = \frac{1}{ic}T_{4k} = -\frac{1}{2}\left(\dot{\psi}\frac{\partial\psi}{\partial x_k} + \frac{\partial\psi}{\partial x_k}\dot{\psi}\right) = -\frac{1}{2}\left(\pi\frac{\partial\psi}{\partial x_k} + \frac{\partial\psi}{\partial x_k}\pi\right) \tag{6.10}$$

which is Hermitian since $(\pi \,.\, \partial\psi/\partial x_k)^* = \partial\psi^*/\partial x_k \cdot \pi^* = \partial\psi/\partial x_k \cdot \pi$. The energy density $-T_{44}$ agrees with H (6.4). The symmetry condition $T_{\mu\nu} = T_{\nu\mu}$ is evidently satisfied with (6.9). In order to verify the validity of the continuity equation (2.13), we write, for example:[1]

$$\dot{G}_k = -\frac{1}{2}\left\{\dot{\pi}\frac{\partial\psi}{\partial x_k} + \frac{\partial\psi}{\partial x_k}\dot{\pi} + \pi\frac{\partial\dot{\psi}}{\partial x_k} + \frac{\partial\dot{\psi}}{\partial x_k}\pi\right\},$$

for which we obtain on account of (6.5 and 6.6):

$$\dot{G}_k = -\left\{\frac{\partial\psi}{\partial x_k}\cdot c^2(\nabla^2 - \mu^2)\psi + \dot{\psi}\frac{\partial\dot{\psi}}{\partial x_k}\right\},$$

[1] One may notice that $[H, ab] = [H, a]b + a[H, b]$.

By a short calculation this is seen to agree with $-\sum_{l=1}^{3} \partial T_{lk}/\partial x_l$. It follows likewise that $\dot{H} = -\nabla \cdot S$. With this equation all requirements for the energy-momentum tensor are satisfied.

We shall now determine the stationary states of the system, defined by (6.4, 6.5), by applying the well known quantum mechanical methods. Although the integral Hamiltonian function $H = \int dx\, H(x)$ is formally constructed by adding the contributions of the single volume elements dx, one still could not say that the variables $(q_j = \psi_\sigma^{(s)},\ p_j = \delta x^{(s)} \cdot \pi_\sigma^{(s)})$ in H are separated, since $dx \cdot |\nabla\psi|^2$ depends not only on the value of the ψ-function in dx, but also on its values in adjoining volume elements. We could obtain the separation, however, by transition to new canonical variables q, p with the help of a spatial Fourier decomposition of $\psi(x)$ and $\pi(x)$ according to (5.1):

$$\psi(x) = V^{-1/2} \sum_k q_k e^{ikx}, \quad \pi(x) = V^{-1/2} \sum_k p_k e^{-ikx}, \tag{6.11}$$

The operators q_k, p_k are subjected to the Hermiticity condition (5.2) and to the commutation rules (5.4):

$$q_{-k} = q_k^*, \quad p_{-k} = p_k^*; \tag{6.12}$$

$$[q_k, q_{k'}] = [p_k, p_{k'}] = 0, \quad [p_k, q_{k'}] = \frac{h}{i} \delta_{kk'}. \tag{6.13}$$

Indeed, by inserting the Fourier series (6.11) into (6.4), it follows:

$$H(x) = \frac{1}{2} \frac{1}{V} \sum_k \sum_{k'} \left\{ p_k p_{k'} e^{-i(k+k')x} + c^2(-k k' + \mu^2) q_k q_{k'} e^{i(k+k')x} \right\};$$

After integration over the periodicity cube V (cf. §5), only the terms $k = -k'$ remain in the double sum, and one obtains with the help of (6.12) for the integral Hamiltonian function:

$$H = \int_V dx\, H = \frac{1}{2} \sum_k \{ p_k^* p_k + \omega_k^2 q_k^* q_k \}, \tag{6.14}$$

where:

$$\omega_k = c\sqrt{\mu^2 + k^2}\ (> 0). \tag{6.15}$$

Here H is now separated into the contributions of the individual lattice points in k-space.

One can obtain a representation of the operators q_k, p_k by matrices which satisfy the conditions (6.12 and 6.13) and make the Hamiltonian (6.14) diagonal, by the following construction: With each degree of freedom k we associate a quantum number N_k, which can assume all integral, non-negative values: $N_k = 0, 1, 2, \ldots$, and we form the two matrices which are Hermitian conjugate to each other:

$$(6.16) \qquad a_k = \begin{pmatrix} 0 & \sqrt{1} & 0 & 0 & \ldots \\ 0 & 0 & \sqrt{2} & 0 & \ldots \\ 0 & 0 & 0 & \sqrt{3} & \ldots \\ 0 & 0 & 0 & 0 & \ldots \\ \ldots & \ldots & \ldots & \ldots & \ldots \end{pmatrix}, \qquad a_k^* = \begin{pmatrix} 0 & 0 & 0 & 0 & \ldots \\ \sqrt{1} & 0 & 0 & 0 & \ldots \\ 0 & \sqrt{2} & 0 & 0 & \ldots \\ 0 & 0 & \sqrt{3} & 0 & \ldots \\ \ldots & \ldots & \ldots & \ldots & \ldots \end{pmatrix}$$

One may write for their matrix elements:

$$(a_k)_{N_k', N_k''} = (a_k^*)_{N_k'', N_k'} = \sqrt{N_k''} \cdot \delta_{N_k', N_k''-1}.$$

With regard to the other quantum numbers N_l $(1 \neq k)$, a_k shall operate as a unit matrix. Written in detail the matrix elements are:

$$(6.17) \quad (a_k)_{N_1' N_2' \ldots, N_1'' N_2'' \ldots} = (a_k^*)_{N_1'' N_2'' \ldots, N_1' N_2' \ldots} = \sqrt{N_k''}\, \delta_{N_k', N_k''-1} \cdot \prod_{l \neq k} \delta_{N_l', N_l''}.$$

According to the known rules of matrix multiplication one finds easily:

$$(6.18) \quad \begin{cases} (a_k a_k^*)_{N_1' N_2' \ldots, N_1'' N_2'' \ldots} = (N_k' + 1) \cdot \prod_l \delta_{N_l', N_l''}, \\ (a_k^* a_k)_{N_1' N_2' \ldots, N_1'' N_2'' \ldots} = N_k' \cdot \prod_l \delta_{N_l', N_l''}. \end{cases}$$

$a_k a_k^*$ and $a_k^* a_k$ are thus diagonal matrices with integral elements. Their difference is the unit matrix: $[a_k, a_k^*] = 1$. We obtain, therefore:

$$(6.19) \qquad [a_k, a_{k'}] = [a_k^*, a_{k'}^*] = 0, \quad [a_k, a_{k'}^*] = \delta_{kk'}.$$

If we choose now for q_k, p_k the following matrices:

$$(6.20) \qquad q_k = \sqrt{\frac{h}{2\omega_k}}\,(a_k + a_{-k}^*), \quad p_k = \sqrt{\frac{h\omega_k}{2}}\, i\,(a_k^* - a_{-k}),$$

then all conditions (6.12) will evidently be satisfied, and the commutation rules (6.13) follow immediately from (6.19) (in particular also for $k' = -k$). Substituting (6.20) into (6.14), we have (with $\omega_{-k} = \omega_k$):

(6.21)
$$H = \frac{h}{2} \sum_k \omega_k (a_k^* a_k + a_k a_k^*).$$

According to (6.18), the energy matrix is diagonal, and its elements, the energy eigen values, are:

(6.22)
$$H_{N_1 N_2 \ldots} = \frac{h}{2} \sum_k \omega_k (2 N_k + 1).$$

The lowest eigen value is obtained by putting all $N_k = 0$:

(6.23)
$$H_0 = \frac{h}{2} \sum_k \omega_k.$$

H_0 is called the zero point energy of the field. It is infinite, i.e., the sum (6.23) is divergent. Since it is an additive constant to the energy, H_0 should not have any physical significance.

After subtracting the zero point energy, there remains for the energy eigen values (6.22):

(6.24)
$$H_{N_1 N_2 \ldots} - H_0 = \sum_k N_k \cdot h \omega_k.$$

The stationary state of the field, which is characterized by the quantum numbers $N_1, N_2, \ldots$, has therefore the same energy (per periodicity domain V) as if there existed in V N_1 corpuscles of the energy $h\omega_1$, N_2 corpuscles of the energy $h\omega_2$, etc. But according to (6.15) $h\omega_k$ is equal to the energy which, according to relativistic mechanics, is attributed to a corpuscle with the rest mass $m = h\,\mu/c$, and the momentum $p = hk$:

(6.25)
$$h\,\omega_k = h\,c\sqrt{\mu^2 + k^2} = c\sqrt{m^2 c^2 + p^2}.$$

Here we see for the first time how the quantization reveals the corpuscular properties of a system described by the field function ψ: the integer N_k represents the number of corpuscles present with the momentum hk.

This interpretation can be corroborated by computing the field momentum $G = \int_V dx\, \mathsf{G}$. The operator of the momentum density, according to (6.10), is:

$$\mathsf{G} = -\frac{1}{2}(\pi \nabla \psi + \nabla \psi \pi).$$

If one substitutes into this the Fourier series (6.11) and if one integrates over the periodicity cube V, one obtains for the total momentum of the domain V:

(6.26) $$G = -\frac{i}{2}\sum_k k\,(q_k p_k + p_k q_k).$$

In this expression we write q_k and p_k with the help of (6.20) again in terms of the matrices (6.17). For symmetry reasons:

$$\sum_k k\,(a^*_{-k} a_{-k} + a_{-k} a^*_{-k}) = -\sum_k k\,(a^*_k a_k + a_k a^*_k),$$

$$\sum_k k\,(a_{-k} a_k + a_k a_{-k}) = 0, \qquad \sum_k k\,(a^*_{-k} a^*_k + a^*_k a^*_{-k}) = 0$$

and one obtains:

(6.27) $$G = \frac{h}{2}\sum_k k\,(a^*_k a_k + a_k a^*_k).$$

According to (6.18), G appears now as a diagonal matrix; its commutability with the diagonal matrix H corresponds to the theorem of the conservation of momentum. The eigen values of the momentum are according to (6.18):

$$G_{N_1 N_2 \ldots} = \frac{h}{2}\sum_k k\,(2N_k + 1);$$

The "zero point momentum" $h/2 . \sum_k k$ vanishes for reasons of symmetry (each pair of vectors with k and $-k$ cancel in the sum). One can therefore also write:

(6.28) $$G_{N_1 N_2 \ldots} = \sum_k N_k \cdot h\,k.$$

This corresponds again to the corpuscular interpretation: in the stationary state characterized by N_1, N_2, ... N_k particles exist with the momentum hk.

Since all energy eigen values (6.22) are single, each stationary state of the total system has the same statistical weight. This is the weight function which characterizes Bose-Einstein statistics.[1] On account of the quantization with

[1] In contradistinction to the Boltzman statistics which ascribes the weight:

$$\frac{\left(\sum_k N_k\right)!}{\prod_k N_k!}$$

to the state in which the kth cell is occupied $N_k$fold.

the canonical commutation relation (1.7), our scalar field thus describes a system of Bose-Einstein particles. We shall see later on that this result for the statistics is not restricted to the scalar field nor is it dependent on the special choice of the field equation (6.1). It has its roots rather in the commutation rules (1.7), and one speaks therefore also of the "field quantization according to Bose-Einstein statistics."[1]

The parameter μ in the above formulas can assume any (real) value. In particular μ may be chosen zero. In that case the rest mass of the corresponding particle vanishes. ($h\omega_k = c \cdot h|k|$) Particles with non-vanishing rest mass and with integral spin are usually called "mesons" on account of the supposed connection with the "mesotrons" of the cosmic radiation. The corresponding ψ-fields are then also called "meson fields." The particular case just discussed refers to the real scalar meson field.

§ 7. Real Field with Sources

Our equations so far describe only the free motions of the "mesons": the occupation numbers N_k are constant with respect to the time. Each particle retains its momentum, and no new particles are either created or annihilated. We shall generalize the theory so that this restriction no longer holds. We shall take as a model the theory of light. It is well known that one describes the expansion of *light waves* in vacuum by homogeneous wave equations. In order to introduce the interaction with matter which is responsible for the emission and absorption of light, one changes from the homogeneous to the inhomogeneous wave equations. Hence, in order to describe in an analogous manner the emission and absorption of the scalar meson field, we must change from the homogeneous Schrödinger-Gordon equation (6.1) to an inhomogeneous wave equation:

$$(\Box - \mu^2)\,\psi = \frac{1}{c}\,\eta\,(x, t). \tag{7.1}$$

Corresponding to this, the modified Lagrangian is:

$$L = L^0 - c\,\eta\,\psi, \text{ where } L^0 = -\frac{c^2}{2}\left\{\sum_{\nu}\left(\frac{\partial \psi}{\partial x_\nu}\right)^2 + \mu^2\psi^2\right\}. \tag{7.2}$$

Here the function η as well as ψ must be Lorentz-invariant. In analogy to

[1] A counterpart to this is the quantization according to Fermi-Dirac statistics which must be applied to the electron wave field; cf. §§20 ff.

optics, we assume the η is connected with the density of matter, which interacts with the ψ-field. If the interacting particles are also described by a quantized field Ψ, then η must be written as an invariant function of the Ψ-components and their derivatives (for instance, $\eta = \text{const.}\ \Psi^2$ for a real scalar field Ψ). Naturally there will be a certain reaction on the Ψ-field by the ψ-field. In order to include that in the theory, one must add in (7.2) the Lagrangian of the corresponding "vacuum field" Ψ. This side of the problem shall be disregarded for the time being. Here we shall consider the behavior of the outside field as given and consider η as a given space-time function. This procedure is consistent if the Ψ-particles are much heavier than the x-particles (i. e., mesons of mass $m = h\mu/c$). In this case, processes of emission and absorption of mesons will produce in the heavy particles only slight recoils which, in first approximation, may be neglected. We are thinking here of the interaction of mesons with protons or neutrons. The masses of the latter are large compared with the masses of the cosmic ray mesotrons. If in first approximation the protons and neutrons are considered infinitely heavy, they will in no way be influenced by the mesons, while these are influenced in accordance with the equations (7.1, 7.2). In the special case where the protons (neutrons) are at rest, then, if their space coordinates x_n ($n = 1, 2 \ldots$) are known, the density function η may be written with the help of the δ-function [cf. (1.8)] as follows:

$$\eta = \sum_n g_n\, \delta(x - x_n), \qquad g_n = \text{const. (real)}. \tag{7.3}$$

Here it is assumed that the meson-proton interaction is a contact interaction. ($\eta \neq 0$ only for $x = x_n$.) We shall interpret all interaction over a finite distance as indirect effects, caused by fields, in order to be sure that their expansion velocity cannot exceed the velocity of light. If the heavy particles interact in the same way with the mesons, then the g_n in (7.3) is independent of n and we have the special case:

$$\eta = g \sum_n \delta(x - x_n). \tag{7.4}$$

The parameter g, as can easily be seen, has the dimension of an electric charge.

The system characterized by the Lagrangian (7.2) shall now be quantized according to the canonical rules. The additional term $-c\eta\psi$ in L does not alter the definition of the momentum π which is canonically conjugate to ψ [cf. (6.3)]. One obtains therefore for the Hamiltonian instead of (6.4):

$$(7.5)\qquad \begin{cases} H = H^0 + H', \\ H^0 = \frac{1}{2}\{\pi^2 + c^2|\nabla\psi|^2 + c^2\mu^2\psi^2\}, \\ H' = c\,\eta\,\psi. \end{cases}$$

On account of the commutation rules (6.5) one finds instead of (6.6) the canonical field equations:

$$(7.6)\qquad \dot{\psi} = \pi, \quad \dot{\pi} = c^2(\nabla^2 - \mu^2)\psi - c\,\eta;$$

The elimination of π from these equations leads back to the inhomogeneous wave equation (7.1):

$$-\frac{1}{c^2}\ddot{\psi} + (\nabla^2 - \mu^2)\psi = \frac{1}{c}\eta.$$

We shall examine the relativistic invariance of the quantization method, assuming that η be a given invariant space-time function. Since H might depend on the time t explicitly, the formulas (4.4 ff.) with $E = H$ can, in general, not be used [cf. (4.5)].[1] For this reason we go back to the original definition of the time-dependent operators $\psi(x, t)$, $\pi(x, t)$, according to which these operators obey as space-time functions the canonical field equations of the classical field which are in analogy to (7.6):

$$\frac{\partial\psi(x,t)}{\partial t} = \pi(x,t), \qquad \frac{\partial\pi(x,t)}{\partial t} = c^2(\nabla^2 - \mu^2)\psi(x,t) - c\,\eta(x,t),$$

In addition to that we require: $\psi(x, 0) = \psi(x)$, $\pi(x, 0) = \pi(x)$. We expand η in powers of t:

$$\eta(x,t) = \sum_n \frac{1}{n!} t^n \overset{(n)}{\eta}(x),$$

The power series of the operators $\psi(x, t)$, $\pi(x, t)$:

$$\psi(x,t) = \sum_n \frac{1}{n!} t^n \overset{(n)}{\psi}(x), \quad \pi(x,t) = \sum_n \frac{1}{n!} t^n \overset{(n+1)}{\psi}(x)$$

[1] There exists, however, the possibility of putting in (4.4) $E = H^0 = \int dx\, H^0$, so that $\psi(x, t)$ obeys the homogeneous wave equation (4.17) and the invariant commutation rules (6.8). This possibility is of interest in view of a "multiple-time" theory of the meson-proton interaction. We shall not enter into this, but refer to §18, where the multiple-time formalism will be explained with the example of the light-electron interaction. (Cf. also Stueckelberg, *Helv. Phys. Acta* 11, 225, 1938.)

may be determined by successive calculation of the coefficients with the help of the recursion formula:

$$\overset{(n+2)}{\psi}(x) = c^2(\nabla^2 - \mu^2)\overset{(n)}{\psi}(x) - c\overset{(n)}{\eta}(x),$$

where $\overset{(0)}{\psi}(x) = \psi(x)$, $\overset{(1)}{\psi}(x) = \pi(x)$. The result can be represented in the following form:

$$\overset{(2m)}{\psi}(x) = [c^2(\nabla^2 - \mu^2)]^m \psi(x) + \sum_k d_{mk} \overset{(k)}{\eta}(x),$$

$$\overset{(2m+1)}{\psi}(x) = [c^2(\nabla^2 - \mu^2)]^m \pi(x) + \sum_k d'_{mk} \overset{(k)}{\eta}(x),$$

where d_{mk}, d'_{mk} are certain differential operators. In the commutators:

$$[\psi(x,t), \psi(x',t')] = \sum_{n,n'} \frac{1}{n!}\frac{1}{n'!} t^n t'^{n'} \left[\overset{(n)}{\psi}(x), \overset{(n')}{\psi}(x')\right]$$

etc., the $\overset{(k)}{\eta}$-terms do not contribute anything, since they commute with $\psi(x)$ and $\pi(x)$. Hence one obtains the same invariant commutation rules (6.8) as in the case of the vacuum field ($\eta = 0$).

If one forms the energy-momentum tensor according to the canonical rules (2.8), one is led back to equation (6.9) where L is replaced by $L^{(0)} - c\eta\psi$. Instead of the continuity equations (2.13), one finds by a short calculation:[1]

$$\dot{H} + \nabla \cdot S = c\psi \frac{\partial \eta}{\partial t}, \quad \dot{G}_k + \sum_j \frac{\partial T_{jk}}{\partial x_j} = -c\psi \frac{\partial \eta}{\partial x_k}$$

These equations formally agree with the classical equations (2.11). The right hand sides of these equations represent the sources of energy and momentum which describe the energy-momentum transfer from the protons to the meson field. It is seen that for the special choice (7.3) for η, there is only momentum—but no energy—exchange, corresponding to the fact that the energy of a stationary proton is constant.

---

[1] In case that $\frac{\partial \eta}{\partial t} \neq 0$, one has obviously:

$$\dot{H} = \frac{i}{h}[H, H] + \frac{\partial H}{\partial t}$$

We carry out the transition to momentum space, as in §6, with the help of the Fourier series (6.11). The coefficients of this series satisfy the conditions (6.12 and 6.13). For the Hamiltonian operator H^0 of the vacuum field [cf. (7.5)] one finds again the expression (6.14):

$$H^0 = \int dx\, H^0 = \frac{1}{2} \sum_k \{p_k^* p_k + \omega_k^2 q_k^* q_k\}.$$

For the calculation of the interaction operator H' we choose the special form (7.4):

$$(7.7) \quad \begin{cases} H' = \int dx\, H' = g\,c \sum_n \int dx\, \delta(x - x_n)\, \psi(x) = g\,c \sum_n \psi(x_n) \\ \quad = g\,c\,V^{-1/2} \sum_k q_k \sum_n e^{i k x_n}. \end{cases}$$

In the special matrix representation (6.20, 17) for q_k and p_k, H^0 will be identical with the diagonal matrix (6.22):

$$(7.8) \qquad H^0_{N_1 N_2 \ldots} = h \sum_k \omega_k \left(N_k + \frac{1}{2}\right),$$

while H' will be a non-diagonal matrix:

$$(7.9) \quad \begin{cases} H' = g\,c \sqrt{\dfrac{h}{2V}} \displaystyle\sum_k \frac{1}{\sqrt{\omega_k}} (a_k + a_{-k}^*) \sum_n e^{i k x_n} \\ \quad = g\,c \sqrt{\dfrac{h}{2V}} \displaystyle\sum_k \frac{1}{\sqrt{\omega_k}} \left(a_k \sum_n e^{i k x_n} + a_k^* \sum_n e^{-i k x_n}\right). \end{cases}$$

For a sufficiently small coupling parameter g, H' may be considered as a small perturbation of the unperturbed vacuum field. We characterize the stationary states of the vacuum field as in §6 by the quantum numbers N_k, which indicate how many mesons with momentum hk are present. The "switching on" of the perturbation H' will then cause transitions between the states of the unperturbed system. On account of the selection rules for the perturbation matrix [cf. (7.9) and (6.17)] the non-vanishing matrix elements refer to transitions in which only one of the occupation numbers N_μ changes by ± 1, i.e., these transitions consist of meson emission and absorption processes only. These transitions can actually take place as real processes only if conservation law of energy is satisfied, i.e., the emitting protons or neutrons must emit or absorb energies of the amount $h\omega_k = hc\sqrt{\mu^2 + k^2}\,(>hc\mu)$. This is only possible if they are under the influence of outside forces, for instance, if they are bound inside a nucleus. We shall not consider such processes here because they involve the wave nature of the heavy particles. The simple expression (7.3 or 7.4) is not adequate for this problem. It is known however that the transitions described by H' can play a role as transitions to "virtual intermediate states" in a second or higher order perturbation theory even if the energy of these states is different from the energy of the initial state. We shall discuss two examples.

We introduce here the perturbation matrix of the second approximation:

$$H''_{\mathrm{F\,I}} = \sum_{\mathrm{II}} \frac{H'_{\mathrm{F\,II}}\, H'_{\mathrm{II\,I}}}{H^0_{\mathrm{I}} - H^0_{\mathrm{II}}}; \tag{7.10}$$

The indices I, II, and F refer to initial, intermediate, and final state. H^0_{I}, H^0_{II} are the corresponding energy eigen values of the unperturbed system, while $H'_{\mathrm{II\,I}}$ and $H'_{\mathrm{F\,II}}$ represent the matrix elements of the perturbation function H' for the virtual transitions I $\rightarrow$ II and II $\rightarrow$ F. If none of the energy denoninators in (7.10) vanish (or become small), then, as is well known, the matrix H'' describes in the first order approximation the effects which are quadratic in H'.

We assume that in the initial state, there exists a proton and a meson with momentum hk ($N_k = 1$, $N_l = 0$ for $l \neq k$). The two-step process, absorption of the meson and emission of another one with momentum hk', corresponds to a scattering of the meson by the proton. Since the proton was assumed to be infinitely heavy this is an elastic scattering. No energy is transferred to the proton and we have $\omega_{k'} = \omega_k$, $|k'| = |k|$. The difference of energy in the initial and intermediate state is:

$$H^0_{\mathrm{I}} - H^0_{\mathrm{II}} = h\,\omega_k.$$

Since according to (6.17) a_k describes the absorption and a_k^* the emission processes, one obtains with the help of (7.9) for the transitions I $\rightarrow$ II and II $\rightarrow$ F the matrix elements:

$$H'_{\mathrm{II\,I}} = g\,c\sqrt{\frac{h}{2V\omega_k}}\,e^{ikx_1}, \quad H'_{\mathrm{F\,II}} = g\,c\sqrt{\frac{h}{2V\omega_{k'}}}\,e^{-ik'x_1}.$$

There exists another way, or rather a different intermediate state II$'$ for the same scattering process: the proton can first emit the meson k' and then absorb the meson k, i.e., the two partial processes can change their sequence:

$$H'_{\mathrm{II'\,I}} = g\,c\sqrt{\frac{h}{2\,V\,\omega_{k'}}}\,e^{-ik'x_1} = H'_{\mathrm{F\,II}},\; H'_{\mathrm{F\,II'}} = g\,c\sqrt{\frac{h}{2\,V\,\omega_k}}\,e^{ikx_1} = H'_{\mathrm{II\,I}}.$$

Since the intermediate state contains two mesons with the same energy $h\omega_k$, the energy denominator is in this case:

$$H^0_{\mathrm{I}} - H^0_{\mathrm{II'}} = -\,h\,\omega_k.$$

Summing up over the two intermediate states according to (7.10), we find:

$$H''_{\mathrm{F\,I}} = \frac{H'_{\mathrm{F\,II}}\,H'_{\mathrm{II\,I}}}{H^0_{\mathrm{I}} - H^0_{\mathrm{II}}} + \frac{H'_{\mathrm{F\,II'}}\,H_{\mathrm{II'\,I}}}{H^0_{\mathrm{I}} - H^0_{\mathrm{II'}}}$$

$$= \frac{H'_{\mathrm{F\,II}}\;H'_{\mathrm{II\,I}}}{h\omega_k} + \frac{H'_{\mathrm{II\,I}}\,H'_{\mathrm{F\,II}}}{-\,h\omega_k} = 0.$$

The contributions of both paths cancel exactly, i.e., the scattering vanishes in this approximation. Only a consideration of the recoils which the proton undergoes during the virtual emission and absorption processes, would lead to a non-vanishing scattering probability.

Another effect, which can be described with the second approximation of the perturbation theory, is the *interaction between two heavy particles (protons), brought about by the meson field.* The energy eigen values of the perturbed system are determined in second approximation, as is well known, by the diagonal elements of the matrix H'' (7.10):

$$H_{\mathrm{I}} = H^0_{\mathrm{I}} + H''_{\mathrm{I\,I}}.\text{[1]}$$

We calculate the eigen value perturbation for the ground state of the vacuum field (all $N_k = 0$):

$$H''_{\mathrm{I_0\,I_0}} = -\sum_{\mathrm{II}} \frac{H'_{\mathrm{I_0\,II}}\;H'_{\mathrm{II\,I_0}}}{H^0_{\mathrm{II}} - H^0_{\mathrm{I_0}}}. \tag{7.11}$$

The virtual transitions $\mathrm{I}_0 \rightarrow \mathrm{II} \rightarrow \mathrm{I}_0$, which contribute to (7.11), consist in the emission of a meson k by a proton n and its absorption by another proton n' (possibly also by the same proton $n' = n$). One obtains for the energy denominator and the matrix element H' according to (7.8 and 7.9):

$$H^0_{\mathrm{II}} - H^0_{\mathrm{I_0}} = h\,\omega_k,$$

$$H'_{\mathrm{II\,I_0}} = g\,c\sqrt{\frac{h}{2\,V\,\omega_k}}\sum_n e^{-ikx_n},\quad H'_{\mathrm{I_0\,II}} = g\,c\sqrt{\frac{h}{2\,V\,\omega_k}}\sum_{n'} e^{ikx_{n'}}.$$

[1] The diagonal elements of H' (7.9) are zero.

Consequently (7.11) yields:

$$(7.12)\qquad \begin{cases} H''_{I_0 I_0} = -\frac{1}{2} g^2 c^2 \sum_n \sum_{n'} \frac{1}{V} \sum_k \frac{e^{ik(x_{n'}-x_n)}}{\omega_k^2} \\ \quad = -\frac{1}{2} g^2 \sum_n \sum_{n'} U(x_{n'} - x_n), \end{cases}$$

where the space function $U(x)$ is defined by:

$$(7.13)\qquad U(x) = c^2 \frac{1}{V} \sum_k \frac{e^{ikx}}{\omega_k^2} = \frac{1}{V} \sum_k \frac{e^{ikx}}{\mu^2 + k^2} = U(-x)$$

[Cf. (6.15)]. The energy of the system in the ground state is thus a function of the relative coordinates of the protons. For each proton pair (n, n') there results a potential energy $-g^2 U(x_{n'} - x_n)$.

In order to change in (7.13) the summation over the k-lattice points (cf. §5) into an integration over the continuous k-space, we notice a volume element dk contains (for the limit $V \to \infty$), $V \cdot (2\pi)^{-3} dk$ lattice points, i.e., $V^{-1}\sum_k \ldots$ has to be replaced by $(2\pi)^{-3}\int dk \ldots$:

$$(7.14)\qquad U(x) = \frac{1}{(2\pi)^3} \int dk \frac{e^{ikx}}{\mu^2 + k^2}.$$

Introducing polar coordinates in k-space in such a way that the axis of the polar coordinate system is parallel to the vector x, and denoting with ϑ the angle between k-vector and the polar axis ∂, we have $dk = 2\pi \sin\vartheta \, d\vartheta \cdot \kappa^2 d\kappa$ $(\kappa = |k|)$:

$$U(x) = \frac{1}{(2\pi)^2} \int_0^\infty d\varkappa \frac{\varkappa^2}{\mu^2 + \varkappa^2} \int_0^\pi d\vartheta \sin\vartheta \, e^{i|x|\varkappa\cos\vartheta}$$

$$= \frac{1}{(2\pi)^2 i |x|} \int_0^\infty d\varkappa \frac{\varkappa}{\mu^2 + \varkappa^2} (e^{i|x|\varkappa} - e^{-i|x|\varkappa})$$

$$= \frac{1}{(2\pi)^2 i |x|} \int_{-\infty}^{+\infty} d\varkappa \frac{\varkappa}{\mu^2 + \varkappa^2} e^{i|x|\varkappa}$$

$$= \frac{1}{2(2\pi)^2 i |x|} \int_{-\infty}^{+\infty} d\varkappa \left(\frac{1}{\varkappa - i\mu} + \frac{1}{\varkappa + i\mu} \right) e^{i|x|\varkappa}.$$

The path of integration can be displaced into the positive-imaginary half of the complex κ-plane such that only the residue at the pole $\kappa = + i\mu$ contributes anything to the integral:

$$U(x) = \frac{1}{4\pi} \frac{e^{-\mu |x|}}{|x|}. \tag{7.15}$$

One easily verifies that this function satisfies the differential equation $\nabla^2 U = \mu^2 U$. More precisely, $(\nabla^2 - \mu^2) U$ vanishes for $x \neq 0$. One obtains from (7.14) [cf. also (4.24)] the more general formula:

$$(-\nabla^2 + \mu^2)\, U(x) = \frac{1}{(2\pi)^3} \int dk\, e^{ikx} = \delta(x). \tag{7.16}$$

Instead of carrying out the k-integration in (7.14), we could also have derived the formula (7.15) for $U(x)$ by integrating the differential equation (7.16).

The space function (7.15) shall be called the "Yukawa potential function" after the Japanese physicist Yukawa who was the first to suggest that the nuclear forces could be interpreted as forces transmitted by a scalar field and who introduced the potential function (7.15) (cf. §§9, 14, 15). For small distances ($|x| \ll 1/\mu$) U varies as $|x|^{-1}$, the Coulomb potential. For larger distances, however, U decreases exponentially and thus approaches zero faster than the Coulomb potential. Since the forces are practically ineffective at distances $> 1/\mu$, this distance is called the "range "of the forces. It agrees (up to a factor 2π) with the "Compton wave length" of the interacting mesons ($1/\mu = h/cm$, where m equals the meson mass, cf. §6). The above formulas are also valid for the special case $\mu = 0$, $m = 0$, for in this case the Yukawa potential changes into the Coulomb potential and the "range" becomes infinite. For representing forces with finite range, as the nuclear forces, one needs therefore mesons with non-vanishing rest mass.

By the intermediate action of a neutral meson field two protons thus exert stationary forces on each other with the potential energy:

$$-g^2\, U(x_{n'} - x_n) = -\frac{g^2}{4\pi} \frac{e^{-\mu |x_{n'} - x_n|}}{|x_{n'} - x_n|}; \tag{7.17}$$

the negative sign corresponds to attractive forces. Strictly speaking, the above derivation is valid only if the protons are fixed at the positions x_n; but it can be expected that the potential energy for slow motion of the protons deviates only little from (7.17), as the meson field—in the sense of an adiabatic change—will always correspond approximately to the stationary state for each instant.

This can be tested by including in the theory the motion of the proton. This implies essentially taking into account the conservation of momentum for the virtual meson emission and absorption processes: the emitting proton acquires a recoil momentum $-hk$ and the absorbing proton takes up the meson momentum $+hk$. For the case of two colliding protons with initial momenta P_1, P_2 there exists a certain probability that these protons will have in the final state the momenta $P_1 - hk$, $P_2 + hk$. This probability is easily expressed by the above matrix elements $H'_{I_0\,II}\,H'_{II\,I_0}/(H^0_{II} - H^0_{I_0})$ and it has the same value as in the quantum mechanical treatment of the collision process (in Born's approximation) on the basis of the potential function (7.17), provided that the change of momenta $h|k|$ is sufficiently small. The Yukawa potential (7.17), however, no longer describes the collision processes correctly for changes of momentum $h|k| \gtrsim Mc$ (M = mass of the proton), since in this case one must take into account the energy changes of the protons in the energy denominators $H^0_{II} - H_{I_0}$. This introduces a modification in the Fourier series (7.13) for $U(x)$ for the Fourier coefficients with $|k| \gtrsim Mc/h$, which means a modification of the space dependence of U for the smallest distances ($|x| \lesssim h/Mc$ = Compton wave length of the proton).

In the double sum (7.12), "self-energy terms" ($n = n'$) appear besides the interaction potentials (terms $n \neq n'$). These self-energy terms stem from the fact that each proton can emit a meson and absorb it again. The self-energy of a proton, due to its interaction with the meson field, has according to (7.12) the value $-\frac{1}{2}g^2U(0)$. But the function $U(x)$ at $x = 0$ becomes infinite like $|x|^{-1}$ according to (7.15) (even with proton recoil included). We thus obtain an infinite self-energy of the proton, similar to the infinite electrical self-energy of a point charge [this latter is infinite like $+\frac{1}{2}e^2U(0)$]. One can make the self-energy finite by introducing into the Fourier expansion of H' (7.7, 7.9) a convergence factor (for instance, $e^{-\alpha k^2}$), which "cuts off" the high meson momenta: in that case the Fourier series for $U(x)$ (7.13) converges also for $x = 0$, i.e., $U(0)$ becomes finite. For the representation of H' in x-space [cf. (7.5, .1)]; this means that the δ-functions $\delta(x - x_n)$ in η (7.4) are replaced by regular space functions which in a finite domain of space will be different from zero (for instance by $e^{-\beta(x-x_n)^2}$). This means that one ascribes to the protons a space extension, and a definite density distribution; the convergence factor in the Fourier series has the significance of a "form factor" for the density distribution. However, such a modification of the Hamiltonian necessarily destroys the Lorentz invariance of the theory. Self-energy and self-momentum of a moving proton do not transform like a 4-vector. Such a theory can at best be considered as a meaningful approximation for processes in which the proton velocities remain of a non-relativistic order of magnitude. Later on we shall meet the same difficulty repeatedly; it represents a fundamental defect of the quantum theory of fields in its present form. We shall return to these and related problems later on in a comprehensive way (§23).

The above calculation of the energy of the ground state was based on a second order perturbation theory, in which terms of the third and higher orders in g are neglected. The formula (7.12), however, for the energy perturbation is actually correct in the case of protons at rest, even with respect to higher powers of g. We shall prove this by transforming the Hamiltonian $H = H^0 + H'$ into diagonal form with the unitary matrix S. Let:

$$S = e^{\frac{i g c}{h} V^{-1/2} \sum_k \frac{p_k}{\omega_k^2} \sum_n e^{-i k x_n}} \tag{7.18}$$

One obtains the matrix S^*, Hermitian conjugate to S, by replacing, in S, i by $-i$ and p_k by $p_k^* = p_{-k}$ [cf. (6.12)]; if one writes k instead of $-k$, it follows:

$$S^* = e^{-\frac{i g c}{h} V^{-1/2} \sum_k \frac{p_k}{\omega_k^2} \sum_n e^{-i k x_n}},$$

consequently:

$$S^* S = S S^* = 1, \qquad S^* = S^{-1};$$

Thus S is a unitary operator, as required. The commutation relations (6.13) yield:

$$[q_k, S] = -g c V^{-1/2} \frac{1}{\omega_k^2} \sum_n e^{-i k x_n} \cdot S, \qquad [p_k, S] = 0;$$

hence one obtains by commuting H^0 with S:

$$[H^0, S] = \frac{1}{2} \sum_k \omega_k^2 \, [q_{-k} q_k, S] = \frac{1}{2} \sum_k \omega_k^2 \{q_{-k} [q_k, S] + [q_{-k}, S] q_k\}$$

$$= -\frac{1}{2} g c V^{-1/2} \left\{ \sum_k q_{-k} \sum_n e^{-i k x_n} \cdot S + S \cdot \sum_k q_k \sum_n e^{i k x_n} \right\}$$

$$= -\frac{1}{2} g c V^{-1/2} \sum_k (q_k S + S q_k) \sum_n e^{i k x_n}.$$

If one puts here:

$$q_k S + S q_k = 2 q_k S - [q_k, S]$$

with the above expression for $[q_k, S]$, one obtains by comparison with (7.7 and 7.12):

$$[H^0, S] = -H' \cdot S + S \cdot H''_{I_0 I_0}.$$

Thus the following results:

$$S^{-1} H S = S^{-1} (H^0 + H') S = H^0 + S^{-1}\{[H^0, S] + H' S\}$$
$$= H^0 + H''_{I_0 I_0}$$

H''_{I_0} is the above discussed potential energy of the static proton-proton forces; since it does not contain the q_k and p_k it is simply an additive constant. By transformation with S we have thus written H in the form (H^0 + const.), where H^0 is made diagonal by the matrix representation (6.20). With (7.8) we have for the eigen values of H:

$$H_{N_1 N_2 \ldots} = \sum_k h\,\omega_k \left(N_k + \frac{1}{2}\right) + H''_{I_0 I_0}; \tag{7.19}$$

Thus the energy of the mesons and the energy of the Yukawa potentials for protons at rest are exactly additive. This additivity indicates that the N_k mesons in the various momentum states do not interact with the protons. In particular they can not be scattered by protons. This statement, which was derived above in second approximation of the perturbation theory, is rigorously true, provided the protons are assumed to be infinitely heavy.

§ 8. Complex Vacuum Field [1]

The real field, which was discussed in §§6 and 7, must be considered as electrically neutral, as was proved in §3. Similarly the corresponding corpuscles must be considered as neutral particles, unless a second and similar field exists that could be united with the first according to (3.1) to form a complex field. In order to describe charged mesons also, we thus introduce two real fields ψ_1, ψ_2 or the complex field ψ, ψ^*, constructed from them according to (3.1). We shall deal again with scalar (Lorentz-invariant) field functions, which satisfy the homogeneous Schrödinger-Gordon wave equation *in vacuo*, i.e., in the absence of particles or fields with which they could interact.

$$(\Box - \mu^2)\,\psi = 0, \qquad (\Box - \mu^2)\,\psi^* = 0. \tag{8.1}$$

For the Lagrangian of the vacuum field we choose the invariant expression:

$$L = -c^2 \left\{ \sum_\nu \frac{\partial \psi^*}{\partial x_\nu} \frac{\partial \psi}{\partial x_\nu} + \mu^2 \psi^* \psi \right\} = \tag{8.2}$$
$$= \dot{\psi}^* \dot{\psi} - c^2 (\nabla \psi^* \cdot \nabla \psi) - c^2 \mu^2 \psi^* \psi;$$

If L, according to (3.2), is considered as function of ψ, ψ^* and their derivatives, the field equations (3.3) agree evidently with (8.1). [One could also represent

[1] Pauli and Weisskopf, *Helv. Phys. Acta*, 7, 709, 1934.

L with the help of (3.1) as function of ψ_1, ψ_2 and their derivatives; in this case the fields ψ_1 and ψ_2 contribute additively to L, each contribution being of the form (6.2). The field equations (1.2) then express that ψ_1 and ψ_2 satisfy the Schrödinger-Gordon equation.]

According to (8.2) L is obviously invariant against the gauge transformation (3.10); consequently, according to (3.11) we can assign to the classical field an electrical charge and current density as follows:

$$(8.3) \qquad \varrho = -i\,\varepsilon\,(\dot{\psi}^*\psi - \dot{\psi}\psi^*), \qquad s = i\,\varepsilon\,c^2\,(\psi\,\nabla\,\psi^* - \psi^*\,\nabla\,\psi).$$

One verifies easily that the continuity equation (3.12) is satisfied on account of the wave equation (8.1).

With:

$$(8.4) \qquad \pi = \frac{\partial L}{\partial \dot{\psi}} = \dot{\psi}^*, \quad \pi^* = \frac{\partial L}{\partial \dot{\psi}^*} = \dot{\psi}$$

one obtains for the Hamiltonian according to (3.6):

$$(8.5) \qquad H = \pi^*\pi + c^2\,(\nabla\,\psi^*\cdot\nabla\,\psi) + c^2\mu^2\psi^*\psi.$$

As positive-definite function, H may be interpreted as energy density. We can use the commutation relations (3.9) directly for the quantization of the theory. Thus results:

$$[H(x), \psi(x')] = \pi^*(x)\,[\pi(x), \psi(x')] = \pi^*(x)\cdot\frac{h}{i}\,\delta(x-x'),$$

$$[H(x), \pi(x')] = -c^2\frac{h}{i}\left\{\left(\nabla\,\psi^*(x)\cdot\nabla\,\delta(x-x')\right) + \mu^2\psi^*(x)\,\delta(x-x')\right\},$$

and by integration over the x-space, as in §6:

$$(8.6) \qquad \begin{cases} \dot{\psi}(x) \equiv \dfrac{i}{h}\,[H, \psi(x)] = \pi^*(x), \\[2ex] \dot{\pi}(x) \equiv \dfrac{i}{h}\,[H, \pi(x)] = c^2\left\{\nabla^2\psi^*(x) - \mu^2\psi^*(x)\right\}. \end{cases}$$

One obtains corresponding equations by exchanging quantities with asterisks against those without asterisks. The elimination of π and π^* yields:

$$(8.7) \qquad \begin{cases} \ddot{\psi}(x) = \dot{\pi}^*(x) = c^2\left\{\nabla^2\psi(x) - \mu^2\psi(x)\right\}, \\ \ddot{\psi}^*(x) = \dot{\pi}(x) = c^2\left\{\nabla^2\psi^*(x) - \mu^2\psi^*(x)\right\}, \end{cases}$$

which formally agrees with (8.1).

In order to establish the Lorentz-invariant commutation rules, we define according to (4.4), with $E = H = \int dx\, \mathbf{H}$:

$$\psi(x, t) = e^{\frac{it}{h}H}\, \psi(x)\, e^{-\frac{it}{h}H}, \quad \psi^*(x, t) = e^{\frac{it}{h}H}\, \psi^*(x)\, e^{-\frac{it}{h}H}.$$

Since these operators by (8.7) again satisfy as space-time functions the Schrödinger-Gordon equation; they obey, according to §4, the commutation rules (4.28), where ψ_1 may be identified with ψ and ψ_2 with ψ^*. We know that these relations are equivalent with (3.9). Comparing (4.21) with (8.6) one finds:

$$d_{11}^{(1)} = d_{22}^{(1)} = 0, \quad d_{12}^{(1)} = d_{21}^{(1)} = 1.$$

Thus the commutation rules (4.28) are:

$$(8.8)\quad \begin{cases} [\psi(x, t), \psi(x', t')] = [\psi^*(x, t), \psi^*(x', t')] = 0, \\ [\psi^*(x, t), \psi(x', t')] = [\psi(x, t), \psi^*(x', t')] = \dfrac{h}{i} D(x - x', t - t'); \end{cases}$$

They are obviously invariant. It should be noted that the last two equations are consistent because $D(x, t) = -D(-x, -t)$ [cf. (4.29)].

The Hamiltonian $\mathbf{H}$ (8.5) may be decomposed with the help of (3.1, 3.5) into two additive terms, one of which depends only on ψ_1, π_1 while the other depends only on ψ_2, π_2. Each of these terms separately has the form (6.4) and consequently can be represented as diagonal matrix according to the method given in §6. This representation has, however, the disadvantage that the total electric charge $e = \int dx\, \rho$ is not diagonal, since according to (8.3) ρ and e cannot be decomposed correspondingly. The total charge e is, on account of the continuity equation (3.12), constant in time, i.e., it commutes with H. It must therefore be possible to make H and e simultaneously diagonal. This can be done indeed in the following manner:

We carry out first the Fourier analysis of the non-Hermitian field functions according to (5.5), with the commutation relations (5.6). Then the differential Hamilton function (8.5) becomes:

$$\mathbf{H} = \frac{1}{V} \sum_k \sum_{k'} e^{i(k-k')x} \{p_k^* p_{k'} + c^2 (\mu^2 + k k')\, q_{k'}^* q_k\}.$$

The integration over the periodicity domain V leaves only the terms $k = k'$ of the sum:

(8.9)
$$H = \int_V dx\, \mathsf{H} = \sum_k \{p_k^* p_k + \omega_k^2 q_k^* q_k\},$$

where ω_k is again defined by (6.15). [The Hamiltonian (8.9) is not identical with (6.14), since the conditions (6.12) do not hold here: q_k and q_{-k} are independent of each other.] We shall further compute the total field momentum. The momentum density of the classical theory is determined by (3.7); changing the sequence of the factors, we obtain the Hermitian operator:

$$\boldsymbol{G} = -(\pi \nabla \psi + \nabla \psi^* \pi^*),$$

and from it by inserting the Fourier series and by integrating over V:

(8.10)
$$G = \int_V dx\, \boldsymbol{G} = -i \sum_k k \{p_k q_k - q_k^* p_k^*\}.$$

[For the energy-momentum tensor (3.8) one obtains by suitable arrangement of the sequence of factors the symmetrical tensor:

$$\boldsymbol{T}_{\mu\nu} = c^2 \left(\frac{\partial \psi^*}{\partial x_\mu} \frac{\partial \psi}{\partial x_\nu} + \frac{\partial \psi^*}{\partial x_\nu} \frac{\partial \psi}{\partial x_\mu} \right) + \boldsymbol{L}\, \delta_{\mu\nu},$$

which satisfies the Hermitian condition and the continuity equations (2.13), as can easily be seen.] The electrical charge density is, according to (3.11) or (8.3, 8.4):

$$\varrho = -i\varepsilon(\pi\psi - \pi^*\psi^*) = -i\varepsilon(\psi\pi - \psi^*\pi^*);$$

Both representations are equivalent on account of the commutation rules (3.9). From this there follows also the Hermitian character of the operator ρ.[1] For the total charge we obtain:

(8.11)
$$e = \int_V dx\, \varrho = -i\varepsilon \sum_k \{p_k q_k - p_k^* q_k^*\}.$$

To make energy, momentum, and charge simultaneously diagonal, we put:

(8.12)
$$\begin{cases} q_k = \sqrt{\dfrac{h}{2\omega_k}}\,(a_k + b_k^*), & q_k^* = \sqrt{\dfrac{h}{2\omega_k}}\,(a_k^* + b_k), \\ p_k = \sqrt{\dfrac{h\omega_k}{2}}\, i\,(a_k^* - b_k), & p_k^* = \sqrt{\dfrac{h\omega_k}{2}}\, i\,(-a_k + b_k^*). \end{cases}$$

[1] This representation of ρ has the advantage as compared with others [for instance $\rho = -i\epsilon(\pi\psi - \psi^*\pi^*)$] that it does not yield a "zero point charge" [cf. (8.17)].

Here a_k, a_k^* denote matrices of the type (6.16), and b_k, b_k^* are matrices of the same kind with respect to other quantum numbers, so that the a_k, a_k^* commute with the b_k, b_k^*. Denoting these quantum numbers with $\overset{+}{N}_k$ and $\overset{-}{N}_k$ (they can assume all integral non-negative values), we have:

$$(8.13) \qquad \begin{cases} (a_k)_{\overset{+}{N}'_k, \overset{+}{N}''_k} = (a_k^*)_{\overset{+}{N}''_k, \overset{+}{N}'_k} = \sqrt{\overset{+}{N}''_k} \cdot \delta_{\overset{+}{N}'_k, \overset{+}{N}''_k - 1}, \\ (b_k)_{\overset{-}{N}'_k, \overset{-}{N}''_k} = (b_k^*)_{\overset{-}{N}''_k, \overset{-}{N}'_k} = \sqrt{\overset{-}{N}''_k} \cdot \delta_{\overset{-}{N}'_k, \overset{-}{N}''_k - 1}; \end{cases}$$

regarding all other quantum numbers (not written down), a_k, a_k^*, b_k, b_k^* act as unit matrices. According to (6.18) $a_k a_k^*$, $a_k^* a_k$, $b_k b_k^*$, $b_k^* b_k$ are diagonal matrices with the diagonal elements $\overset{+}{N}_k + 1$, $\overset{+}{N}_k$, $\overset{-}{N}_k + 1$, $\overset{-}{N}_k$; we write down the following eigen values:

$$(8.14) \qquad (a_k^* a_k)_{\overset{+}{N}_1 \overset{+}{N}_2 \dots \overset{-}{N}_1 \overset{-}{N}_2 \dots} = \overset{+}{N}_k, \qquad (b_k^* b_k)_{\overset{+}{N}_1 \overset{+}{N}_2 \dots \overset{-}{N}_1 \overset{-}{N}_2 \dots} = \overset{-}{N}_k.$$

According to (6.19) the following commutation rules hold:

$$(8.15) \qquad [a_k, a_{k'}^*] = [b_k, b_{k'}^*] = \delta_{kk'},$$

while all other pairs of matrices commute. One verifies easily that the matrix representations (8.12) of q_k, q_k^*, p_k, p_k^* satisfy the commutation rules (5.6).

Inserting (8.12) into (8.9 8,.10, 8.11), we obtain:

$$(8.16) \qquad \begin{cases} H = \dfrac{h}{2} \sum\limits_k \omega_k \{a_k^* a_k + a_k a_k^* + b_k^* b_k + b_k b_k^*\}, \\ G = h \sum\limits_k k \{a_k^* a_k - b_k b_k^*\}, \\ e = \dfrac{h\varepsilon}{2} \sum\limits_k \{a_k^* a_k + a_k a_k^* - b_k^* b_k - b_k b_k^*\}. \end{cases}$$

Energy, momentum, and charge are thus diagonal, and their eigen values according to (8.14, 8.15) are:

$$(8.17)\begin{cases} H_{\overset{+}{N}_1 \overset{+}{N}_2 \ldots \overline{N}_1 \overline{N}_2 \ldots} = \frac{h}{2} \sum_k \omega_k \left\{ (2 \overset{+}{N}_k + 1) + (2 \overline{N}_k + 1) \right\} \\ \qquad = \sum_k h\, \omega_k (\overset{+}{N}_k + \overline{N}_k) + 2\, H_0, \\ G_{\overset{+}{N}_1 \overset{+}{N}_2 \ldots \overline{N}_1 \overline{N}_2 \ldots} = h \sum_k k\, (\overset{+}{N}_k - \overline{N}_k - 1) = \sum_k h\, k\, (\overset{+}{N}_k - \overline{N}_k), \\ e_{\overset{+}{N}_1 \overset{+}{N}_2 \ldots \overline{N}_1 \overline{N}_2 \ldots} = \frac{h\varepsilon}{2} \sum_k \left\{ (2 \overset{+}{N}_k + 1) - (2 \overline{N}_k + 1) \right\} = h\, \varepsilon \sum_k (\overset{+}{N}_k - \overline{N}_k). \end{cases}$$

Apart from the infinitely large zero point energy $2\,H_0$ which is twice as large for the complex as for the neutral field [cf. (6.23)], the expressions (8.17) again allow a corpuscular interpretation: if one has particles of mass $m = h\,\mu/c$, with the electric charges $\pm h\epsilon$, and if one assigns to all $\overset{+}{N}_k$ particles of the charge $+h\epsilon$ the momentum $+hk$ and to all $\overline{N}_k$ particles of the charge $-h\epsilon$ the momentum $-hk$, then energy, momentum, and charge of these particles are exactly represented by (8.17). Hence one can state that in the stationary state, which is characterized by the quantum numbers $\overset{+}{N}_1\, \overset{+}{N}_2 \ldots .\, \overline{N}_1\, \overline{N}_2 \ldots$, there always exist $\overset{+}{N}_k$ particles with the momentum hk and the charge $h\epsilon$, as well as $\overline{N}_k$ particles with the momentum $-hk$ and the charge $-h\epsilon$. The quantized complex field represents thus charged "mesons" contrary to the real field, whereby both signs for the charge appear on equal footing. The fact that the charge occurs only in integral multiples of an elementary charge ($h\epsilon$) is in this theory a consequence of the field quantization. Since each eigen value (8.17) must be counted as simple eigen value, the statistical weight 1 must be ascribed to each state $(\overset{+}{N}_1\, \overset{+}{N}_2 \ldots .\, \overline{N}_1\, \overline{N}_2 \ldots)$ of the total system, in agreement with the Bose-Einstein enumeration.

§ 9. Complex Field in Interaction with Protons and Neutrons [1]

In this section the charged meson field shall be coupled with stationary (infinitely heavy) protons and neutrons, as it was done for the neutral field in

[1] Yukawa, *Proc. Phys.-Math. Soc. Japan* 17, 48, 1935.

§7.[1] This can be done formally by adopting the interaction terms (7.1 to 7.5) for the real field components ψ_1, ψ_2 and combining them linearly. Thus one obtains for the Hamiltonian in the complex notation [with $1/\sqrt{2}\,.\,(\eta_1 - i\eta_2) = \eta$]:

$$(9.1)\qquad \begin{cases} H = H^0 + H', \\ H^0 = \pi^* \pi + c^2 \{(\nabla \psi^* \cdot \nabla \psi) + \mu^2 \psi^* \psi\}, \\ H' = c\,(\eta\,\psi + \eta^*\,\psi^*). \end{cases}$$

Nothing essentially new can be said with respect to the Lorentz invariance of the quantized theory as well as the energy-momentum tensor which was discussed in §§7 and 8, so that we shall omit the discussion of these points.

A characteristic difference from the neutral meson field lies in the following. Since according to (5.5) and (8.12) the matrices ψ and ψ^* cause transitions, in which the meson numbers $\overset{+}{N}_k$, $\overset{-}{N}_k$ change by ± 1, the interaction term $H' = \int dx \mathsf{H}'$ describes again processes of emission and absorption of mesons by the protons and neutrons. These mesons carry an electrical charge, and these processes are compatible with the conservation of charge only, if the heavy particles experience corresponding changes of charge by transforming protons and neutrons into each other. For the formal description of such transformations (which also occur in the β-decay) it is convenient, according to Heisenberg,[2] to consider protons and neutrons as two states of one and the same particle, which may be called "proton-neutron" (or also "nucleon"). We characterize the two states by the value of the "charge number" λ: $\lambda = 0$ corresponds to the neutron, $\lambda = 1$ to the proton state. The transition $\lambda = 1 \rightarrow 0$ is connected with the emission of a positive meson or the absorption of a negative meson; the inverse processes are connected with a transition $\lambda = 0 \rightarrow 1$. Naturally the charge of a positive meson must be exactly the same as the charge of the proton (the electrical elementary quantum). For the following consideration it is convenient to express all charges in units of the elementary charge. We therefore write:

$$h\epsilon = 1.$$

Insofar as the density function ρ, introduced in §8, represents an electrical charge density in the usual sense, which together with the proton charges shall

[1] The forces, which the mesons experience on account of their charge by the Coulomb field of the proton, are not considered here. Cf. §11.

[2] *Z. Phys.* 77, 1, 1932.

satisfy a continuity equation, H' must describe proton–neutron transformations simultaneously with the meson emission and absorption processes. This implies that the functions η and η^* in (9.1) must also be matrices with respect to the charge number of the proton-neutrons ($n = 1, 2, \ldots$). In the theory of interaction of charged mesons with nucleons one cannot altogether neglect the changes of states of the nucleons. This is true even if we assume, as in §6, the nucleons to be stationary by ignoring their translatory degrees of freedom. We must still consider their "charge degrees of freedom." The selection rules for the transitions which follow from the conservation of charge will prove to be essential.

For the formulation of this interaction problem, we can again use for η the expression (7.3):

$$\eta = \sum_n g_n \,\delta(x - x_n), \qquad \eta^* = \sum_n g_n^* \,\delta(x - x_n), \tag{9.2}$$

where g_n, g_n^* must now be considered as matrices with regard to the respective charge numbers λ_n. They must be determined in such a way that the total charge is conserved. Let:

$$e^P = \sum_n e_n \tag{9.3}$$

be the charge of all proton-neutrons, represented as a matrix with regard to the λ_n (see below). We now call e^0 the total charge of the meson field:

$$e^0 = \int dx\, \varrho = -\frac{i}{h}\int dx\,(\pi\,\psi - \pi^*\,\psi^*). \tag{9.4}$$

The total charge:

$$e = e^0 + e^P \tag{9.5}$$

must commute with $H = \int dx\, \mathsf{H}$, so that $e = \text{const.}$:

$$[H, e] = [H^0 + H', e] = 0. \tag{9.6}$$

But H^0 commutes with e, since H^0, e^0, and e^P are diagonal matrices with respect to the $\overset{+}{N}_k$, $\overset{-}{N}_k$, and λ_n. Hence we must have:

$$[H', e] = [H', e^0] + [H', e^P] = 0. \tag{9.7}$$

From (9.1, 9.4) follows:

$$[H', e^0] = -\frac{i}{h} \int dx\, ([H', \pi]\, \psi - [H', \pi^*]\, \psi^*);$$

$$[H', \pi(x)] = \int dx'\, [H'(x'), \pi(x)] = -\frac{h}{i}\, c \int dx'\, \eta(x')\, \delta(x - x')$$

$$= -\frac{h}{i}\, c\, \eta(x), \quad [H', \pi^*(x)] = -\frac{h}{i}\, c\, \eta^*(x);$$

hence:

$$[H', e^0] = c \int dx\, (\eta\, \psi - \eta^*\, \psi^*).$$

On the other hand:

$$[H', e^P] = c \int dx\, ([\eta, e^P]\, \psi + [\eta^*, e^P]\, \psi^*).$$

In order to satisfy (9.7) we require:

$$(9.8) \qquad [\eta, e^P] = -\eta, \quad [\eta^*, e^P] = +\eta^*,$$

or according to (9.2, 9.3) [obviously $[g_n, e_{n'}] = 0$ for $n \neq n'$]:

$$(9.9) \qquad [g_n, e_n] = -g_n, \quad [\overset{*}{g}_n, e_n] = +\overset{*}{g}_n.$$

The proton-neutron charge e_n, represented as matrix with respect to the charge number λ_n (= 0, 1), is evidently a diagonal matrix with the diagonal elements 0 (neutron) and 1 (proton):

$$(9.10) \qquad e_n = \begin{pmatrix} 0 & 0 \\ 0 & 1 \end{pmatrix}.$$

With:

$$(9.11) \qquad g_n = g \cdot \begin{pmatrix} 0 & 0 \\ 1 & 0 \end{pmatrix}, \quad \overset{*}{g}_n = g^* \cdot \begin{pmatrix} 0 & 1 \\ 0 & 0 \end{pmatrix},$$

where g is a numerical factor, one finds:

$$g_n\, e_n = 0, \quad e_n\, g_n = g_n, \quad \overset{*}{g}_n\, e_n = \overset{*}{g}_n, \quad e_n\, \overset{*}{g}_n = 0;$$

Thus the conditions (9.9) are satisfied by (9.11). The only non-vanishing elements of the matrices (9.11) are $(g_n)_{1,0}$ and $(\overset{*}{g}_n)_{0,1}$, i.e., g_n describes transitions from the neutron to the proton state, and g_n^* the inverse transformations. For that reason g_n appears in H' multiplied with $\psi(x_n)$, since according to (5.5) and (8.12) ψ describes the absorption of positive, or the emission of negative, mesons.

The Pauli spin-matrices are frequently used for the formal representation of the above matrices:

$$(9.12)\quad \tau^{(1)} = \begin{pmatrix} 0 & 1 \\ 1 & 0 \end{pmatrix}, \qquad \tau^{(2)} = \begin{pmatrix} 0 & -i \\ +i & 0 \end{pmatrix}, \qquad \tau^{(3)} = \begin{pmatrix} 1 & 0 \\ 0 & -1 \end{pmatrix};$$

These of course must be understood as matrices referring to the charge number λ (the "isotopic spin"). The Pauli matrices are Hermitian matrices with the properties:

$$(9.13)\quad \begin{cases} (\tau^{(1)})^2 = (\tau^{(2)})^2 = (\tau^{(3)})^2 = 1, & \tau^{(2)}\tau^{(3)} = -\tau^{(3)}\tau^{(2)} = i\,\tau^{(1)}, \\ \tau^{(3)}\tau^{(1)} = -\tau^{(1)}\tau^{(3)} = i\,\tau^{(2)}, & \tau^{(1)}\tau^{(2)} = -\tau^{(2)}\tau^{(1)} = i\,\tau^{(3)}. \end{cases}$$

With (9.12) we write in place of (9.10, 9.11):

$$(9.14)\qquad e_n = \frac{1}{2}\,(1 - \tau_n^{(3)}),$$

$$(9.15)\qquad g_n = g\cdot\frac{1}{2}\,(\tau_n^{(1)} - i\,\tau_n^{(2)}), \qquad g_n^* = g^*\cdot\frac{1}{2}\,(\tau_n^{(1)} + i\,\tau_n^{(2)});$$

the index n indicates that we have matrices with respect to λ_n, which are unit matrices with respect to the other charge numbers. With the help of the relations (9.13) one can easily verify (9.9).

Summarizing we may write for H', using (9.1, 9.2, 9.15):

$$(9.16)\quad \begin{cases} H' = \int dx\, \mathsf{H}' = c\sum\limits_n \{g_n\,\psi(x_n) + g_n^*\,\psi^*(x_n)\} \\ \quad = c\sum\limits_n \left\{g\cdot\frac{1}{2}\,(\tau_n^{(1)} - i\,\tau_n^{(2)})\,\psi(x_n) + g^*\cdot\frac{1}{2}\,(\tau_n^{(1)} + i\,\tau_n^{(2)})\,\psi^*(x_n)\right\}. \end{cases}$$

Introducing in addition the Fourier series (5.5) and the matrix representation (8.12), it follows:

$$(9.17)\quad \begin{cases} H' = c\sqrt{\frac{h}{2V}}\sum\limits_k \frac{1}{\sqrt{\omega_k}}\left\{(a_k + b_k^*)\sum\limits_n g_n\,e^{ikx_n} + \right. \\ \qquad\qquad \left. + (a_k^* + b_k)\sum\limits_n g_n^*\,e^{-ikx_n}\right\}. \end{cases}$$

If we want to take into account that the masses of proton and neutron are different, we must add to H the rest energies of the nucleons. These are the following diagonal matrices with respect to the charge numbers λ_n:

$$\begin{pmatrix} M_N c^2 & 0 \\ 0 & M_P c^2 \end{pmatrix} = \frac{c^2}{2} \{M_N (1 + \tau_n^{(3)}) + M_P (1 - \tau_n^{(3)})\}.$$

In the following discussion the mass difference $M_N - M_P$ will be considered negligible compared with the mass of the meson.

With H' as perturbation function one can discuss problems like the meson scattering or the nuclear forces with a second order perturbation calculation. It was shown in §7 that the scattering of neutral mesons on proton or neutron at rest vanishes, since the scattering process can take place along two ways which differ in the sequence of the virtual processes (absorption of the primary and emission of the secondary meson). The contributions of these to the matrix element of the scattering process just compensate each other. The situation is quite different for the scattering of charged mesons. For instance a primary positive meson cannot be absorbed by a proton; hence the scattering process cannot take place in such a way that first the primary meson is absorbed and then the secondary meson is emitted. Only the second case is possible, characterized by the following scheme:

$$P + \overset{+}{\mu}_k \rightarrow N + \overset{+}{\mu}_k + \overset{+}{\mu}_{k'} \rightarrow P + \overset{+}{\mu}_{k'}$$

(P = proton, N = neutron, $\overset{+}{\mu}_k$ ($\overset{-}{\mu}_k$) = positive (negative) meson with the momentum hk). Similarly for the scattering of a negative meson on the neutron there exists only the possibility:

$$N + \overset{-}{\mu}_k \rightarrow P + \overset{-}{\mu}_k + \overset{-}{\mu}_{k'} \rightarrow N + \overset{-}{\mu}_{k'}.$$

On the other hand only the absorption of the primary meson occurs as the intermediate process for the scattering of a negative meson on the proton, and for the scattering of the positive meson at the neutron:

$$P + \overset{-}{\mu}_k \rightarrow N \rightarrow P + \overset{-}{\mu}_{k'}, \quad N + \overset{+}{\mu}_k \rightarrow P \rightarrow N + \overset{-}{\mu}_{k'}.$$

In each case the scattering process can thus take place in only one way on account of the conservation of charge. One obtains for the respective matrix element, which cannot be compensated by any other, according to (7.10) and

(9.17) and with $\omega_{k'} = \omega_k$:

$$H''_{\mathrm{F\,I}} = \pm \frac{c^2 |g|^2}{2 V \omega_k^2}.$$

The probability of the scattering can be calculated from this according to well-known methods. The angular distribution of the scattered mesons is obviously isotropic and for the total scattering cross-section one finds:

$$Q = \frac{|g|^4}{4 \pi (h \omega_k)^2}. \tag{9.18}$$

In order to derive the nuclear forces produced by the charged meson field, we compute, as in §7, the perturbation of the energy eigen value by H' for the ground state of the field (all $\overset{+}{N}_k$ and $\overset{-}{N}_k = 0$) to the second order approximation. The virtual transitions $\mathrm{I}_0 \to \mathrm{II} \to \mathrm{I}_0$, which must be considered in (7.11), consist again in the emission of a meson by a proton-neutron n and in its absorption by another proton-neutron n'. Depending on whether a positive or negative meson is emitted, the intermediate state may be indicated by $\overset{+}{\mathrm{II}}$ or $\overset{-}{\mathrm{II}}$. The matrix elements for the virtual partial processes $\mathrm{I}_0 \to \mathrm{II}$ (emission) and $\mathrm{II} \to \mathrm{I}_0$ (absorption) are then according to (9.17) and (8.13):

$$H'_{\overset{+}{\mathrm{II}}\,\mathrm{I}_0} = c \sqrt{\frac{h}{2 V \omega_k}} \sum_n g_n^* e^{-i k x_n}, \quad H'_{\mathrm{I}_0\,\overset{+}{\mathrm{II}}} = c \sqrt{\frac{h}{2 V \omega_k}} \sum_{n'} g_{n'} e^{i k x_{n'}},$$

$$H'_{\overset{-}{\mathrm{II}}\,\mathrm{I}_0} = c \sqrt{\frac{h}{2 V \omega_k}} \sum_n g_n e^{i k x_n}, \quad H'_{\mathrm{I}_0\,\overset{-}{\mathrm{II}}} = c \sqrt{\frac{h}{2 V \omega_k}} \sum_{n'} g_{n'}^* e^{-i k x_{n'}}.$$

In these expressions we have left the g_n which are matrices with respect to the charge number λ_n. It is even permitted to substitute them into $H''_{\mathrm{I}_0\mathrm{I}_0}$(7.11); if the sequence of factors is chosen according to (7.11). With:

$$H^0_{\overset{+}{\mathrm{II}}} - H^0_{\mathrm{I}_0} = H^0_{\overset{-}{\mathrm{II}}} - H^0_{\mathrm{I}_0} = h \omega_k$$

one obtains for $H''_{\mathrm{I}_0\mathrm{I}_0}$:

$$H''_{\mathrm{I}_0\mathrm{I}_0} = -\frac{1}{2} c^2 \frac{1}{V} \sum_k \frac{1}{\omega_k^2} \sum_n \sum_{n'} \{ g_{n'} g_n^* e^{i k (x_{n'} - x_n)} + g_{n'}^* g_n e^{i k (x_n - x_{n'})} \}.$$

Just as it was done in (7.12), we can introduce here the Yukawa potential, defined by (7.13):

$$H''_{I_0 I_0} = -\frac{1}{2} \sum_n \sum_{n'} \{g_{n'} g_n^* + g_{n'}^* g_n\} \, U(x_{n'} - x_n). \tag{9.19}$$

Each term n, n' of this double sum is still a matrix with respect to the proton-neutron charge numbers λ_n, $\lambda_{n'}$, the eigen values of which can easily be determined.

As far as the terms $n = n'$ are concerned, we have, according to (9.11):

$$g_n g_n^* + g_n^* g_n = |g|^2, \tag{9.20}$$

This is the unit matrix multiplied by $|g|^2$. Each of these terms is already diagonal and has the value $-\frac{1}{2}|g|^2 U(0)$. Thus each proton-neutron has a negative-infinite self-energy, on account of its interaction with the charged meson field. Everything that has been said in §7 regarding this problem applies also in this case.

The terms $n \neq n'$ in (9.19) represent forces between pairs of heavy particles. We select the terms $n = 1$, $n' = 2$ and $n = 2$, $n' = 1$:

$$V_{12} = -\{g_1 g_2^* + g_1^* g_2\} \, U(x_1 - x_2). \tag{9.21}$$

The space-dependence of this potential is given by the Yukawa potential (7.15). The matrix factor $\{g_1 g_2^* + g_1^* g_2\}$ indicates that we are dealing here with an "exchange force." The only non-vanishing matrix elements of g_n and of g_n^* correspond according to (9.11) to transitions of the nth particle from the neutron into the proton state, or vice versa. Consequently, the matrix $g_1 g_2^*$ has only one non-vanishing element, corresponding to the initial state:

particle 1 = neutron, particle 2 = proton,

and to the final state:

particle 1 = proton, particle 2 = neutron;

in the matrix $g_1^* g_2$ initial and final states are interchanged, since it is the Hermitian conjugate to $g_1 g_2^*$. If, therefore, the two particles are initially in the same state (two neutrons or two protons), then according to (9.21) there exist no transitions. The reason for this is that two such particles cannot exchange any charged mesons on account of the conservation of charge. If, on the other hand, initially one of the two particles is a neutron, the other a proton, then V_{12} represents a transition to a final state, in which both particles have exchanged their charge. The transfer of charge from one particle to the other is, of course, effected by charged mesons passing from one particle to the other. In case the motion of the heavy particles is being considered in addition to the charge exchange (cf. §7), the mesons cause also an exchange of momentum, according to the scheme:

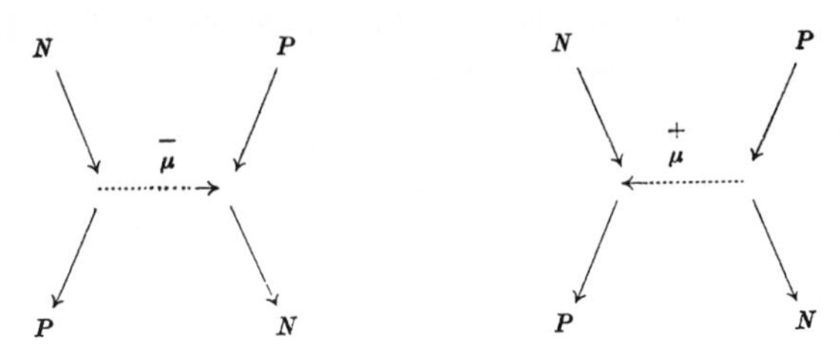

It is characteristic of the exchange force, caused by charged mesons, that the particles interchange their charges in a collison, whereas the neutral meson field (§7) naturally has no influence on their charges. The nuclear forces probably do have partially such an "exchange" character, and it is this fact which caused Yukawa[1] to propose the theory which has been developed here as a theory of the nuclear forces. We shall return later (§15) to such hypothetical relations.

We shall compute the eigen values of $H''_{I_0 I_0}$(9.19) for the case when only two nucleons are present. $H''_{I_0 I_0}$ is then identical with V_{12} (9.21) after subtraction of the self-energies. If we denote the four states of the pair of particles by a, b, c, d in the following sequence: neutron-neutron $= a$, proton-neutron $= b$, neutron-proton $= c$, proton-proton $= d$, then V_{12} can be represented by the following matrix scheme:

	a	b	c	d
a	0	0	0	0
b	0	0	$-\lvert g\rvert^2 U(x_1-x_2)$	0
c	0	$-\lvert g\rvert^2 U(x_1-x_2)$	0	0
d	0	0	0	0

The eigen values of this matrix are evidently:

$$0, \quad -|g|^2\, U(x_1-x_2), \quad +|g|^2\, U(x_1-x_2), \quad 0.$$

The double eigen value 0 belongs to the stationary states a and d; the two others correspond to mixtures of the states b and c; one can see easily that

[1] Cf. footnote 1, p. 53.

indeed the Schrödinger function of the system is symmetrical with respect to the charge numbers λ_1, λ_2 in the case of the eigen value $-|g|^2 U$ and antisymmetrical in case of the eigen value $+|g|^2 U$. In the first case the protons and neutrons attract each other, in the second case they repel each other.

In the literature one finds the exchange operators $(g_{n'} g_n^* + g_{n'}^* g_n)$ usually expressed in terms of the isotopic spin-matrices (9.12). It follows with the help of (9.15):

$$g_{n'} g_n^* + g_{n'}^* g_n = \frac{1}{2} |g|^2 \{\tau_{n'}^{(1)} \tau_n^{(1)} + \tau_{n'}^{(2)} \tau_n^{(2)}\}. \tag{9.22}$$

Instead of (9.21), one can also write:

$$V_{12} = -\frac{1}{2} |g|^2 \{\tau_1^{(1)} \tau_2^{(1)} + \tau_1^{(2)} \tau_2^{(2)}\} U(x_1 - x_2). \tag{9.23}$$

It should be pointed out that in addition to these exchange forces, derived in second approximation of the perturbation theory, there will occur forces in higher approximation, which cause interaction between neutrons as well as between protons. Moreover there will be forces between neutrons and protons without any exchange character. For instance, a neutron can emit a negative meson, followed by a positive meson. These two mesons may then be absorbed in the reverse order by another neutron. This four-step transition and other similar transitions yield in the fourth approximation an ordinary static interaction between the two neutrons. Unfortunately, it is not possible to calculate such higher order interactions in an unambiguous way, on account of the connection of these terms with the self-energy problem. They are only finite if one assigns to protons and neutrons a finite extension in space, which destroys the relativistic invariance. The magnitude of the higher order terms depends not only on the value of the coupling parameter $|g|$, but also on the choice of the proton radius and of the "form factor." Only if one can choose these quantities in such a way that the Yukawa exchange forces (9.23) predominate for particle distances $|x_1 - x_2|$ of the order of magnitude μ^{-1} (= range of forces) do we obtain, to some extent, unambiguous results.

In the case of a neutral meson field interacting with protons and neutrons at rest, we obtained an exact expression of the eigen values of H by transforming H with the unitary matrix S (7.18). So far it has not been possible to find a corresponding solution for the charged meson field, i.e., for the problem (9.1), although it is possible to expand S in powers of $|g|$, which corresponds to the perturbation method used above. No one has yet succeeded, however, in summing up this series in closed form. The formal reason that the problems here are more complicated than in the case of neutral field stems from the non-commuting matrices g_n and g_n^* in H'.

In higher approximations of the perturbation theory there occur also transitions in which two or more mesons are emitted by different protons or neutrons and are absorbed again by others, so that more than two nucleons take part in each process. From such transitions forces will result which depend on the coordinates of three or more nuclear particles (multi-body forces).

If for the dimensionless coupling parameter, $|g|^2/hc \gtrsim 1$, then the expansions of the perturbation theory are useless on account of convergence difficulties.[1] In the limiting case of strong coupling ($|g|^2/hc >> 1$), on the other hand, it is possible again to obtain approximate solutions in the form of development in decreasing powers of $|g|^2$.[2] We cannot deal here with these rather complicated calculations, but we want to mention only one result. According to this theory the nucleon can bind positive or negative mesons, creating proton-isobars with any charge number; the energy or mass of the isobars depends quadratically on the charge number:

$$M_\lambda = A + B\left(\lambda - \frac{1}{2}\right)^2 \qquad (\lambda = \text{integer})$$

One finds for the static forces in lowest approximation (i.e., neglecting terms of higher order in $|g|^{-2}$) and for the limit proton radius $\rightarrow 0$, two-body forces with the potential:

$$V_{12} = -\frac{1}{4}|g|^2 \cdot \mathfrak{P} \cdot U(x_1 - x_2),$$

$\mathfrak{P}$ is an exchange operator, which has a slightly more general significance than the operator (9.21 or 9.23), on account of the existence of higher proton-isobars; this operator changes the charge numbers λ_1, λ_2 of the two particles into $\lambda_1 + 1$, $\lambda_2 - 1$, or into $\lambda_1 - 1$, $\lambda_2 + 1$. Apart from this and from a numerical factor, both forces are the same for strong and weak coupling.

§ 10. Combined Charged and Neutral Field

We shall discuss briefly a generalization of the Yukawa theory, in which the nuclear particles are interacting with a charged as well as with a neutral meson field in a symmetrical way such that the forces become independent of the charge of the nucleons ("symmetrical theory" of Kemmer[3]). We denote the neutral field with ψ_3 ($\psi_3 = \psi_3^*$), and decompose the complex field ψ, according to (3.1) into its real (Hermitian) components ψ_1, ψ_2. For the Hamiltonian we choose:

$$(10.1) \qquad \begin{cases} H = H^0 + H', \\ H^0 = \dfrac{1}{2}\displaystyle\int dx \sum_\sigma \{\pi_\sigma^2 + c^2 |\nabla \psi_\sigma|^2 + c^2\mu^2 \psi_\sigma^2\}, \\ H' = cg' \displaystyle\sum_\sigma \sum_n \tau_n^{(\sigma)} \psi_\sigma(x_n); \end{cases}$$

[1] If one determines $|g|$ by comparison of (9.23) with the strength of the actual nuclear forces (cf. §15), the value for $|g|^2/hc$ will not be much smaller than 1.

[2] Wentzel, *Helv. Phys. Acta 13*, 269, 1940; *14*, 633, 1941. Oppenheimer and Schwinger, *Phys. Rev. 60*, 150, 1941. Serber and Dancoff, *Phys. Rev. 63*, 143, 1943.

[3] Kemmer, *Proc. Cambridge Phil. Soc. 34*, 354, 1938.

The σ-sums are now to be extended over the values 1, 2, 3: $\tau_n^{(1)}$, $\tau_n^{(2)}$, $\tau_n^{(3)}$ are the "isotopic spin-matrices" of the nth nucleon [cf. (9.12)]. The coupling parameter g' is assumed to be real. The terms in (10.1) ($\sigma = 1, 2$) which involve the charged field agree with (9.1) and (9.16), if $g = g^* = g'\sqrt{2}$. With respect to the neutral field ψ_3, there exists a difference as compared with (7.5), since $\psi_3(x_n)$ appears in H' multiplied with $\tau_n^{(3)}$; i.e., the coupling factor g_n in (7.3) has different signs depending on whether the particle n is in the neutron or in the proton state. This has the consequence that $\tau_n^{(\sigma)}$ and $\psi_\sigma(x_n)$ appear combined like the scalar product of two vectors τ_n and $\psi(x_n)$. Of course these are vectors in a symbolic space which has nothing in common with ordinary space (x_1, x_2, x_3).

With the help of the Fourier series (5.1) we have for H^0 and H':

$$(10.2)\qquad \begin{cases} H^0 = \dfrac{1}{2}\sum\limits_k \sum\limits_\sigma \{p^*_{\sigma,k}\, p_{\sigma,k} + \omega_k^2\, q^*_{\sigma,k}\, q_{\sigma,k}\}, \\ H' = c\, g'\, V^{-1/2} \sum\limits_k \sum\limits_\sigma q_{\sigma,k} \sum\limits_n \tau_n^{(\sigma)}\, e^{i k x_n}. \end{cases}$$

The $q_{\sigma,k}$ and $p_{\sigma,k}$ satisfy the Hermitian conditions (5.2) and the commutation rules (5.4). We transform $H = H^0 + H'$ with the help of a matrix S, which we assume to be expanded according to the perturbation method in powers of g':

$$(10.3)\qquad S = 1 + S' + S'' + \dots;$$

we restrict ourselves here to the second approximation and shall therefore suppress all terms which contain third and higher powers of g'. In order that S be unitary, it must be true that:

$$S^*\, S = 1 + (S' + S'^*) + (S'^*\, S' + S'' + S''^*) + \dots = 1.$$

In order to satisfy this condition, we postulate:

$$(10.4)\qquad S'^* = -S'$$

and write:

$$(10.5)\qquad S'' = S''^* = \frac{1}{2}\, S'^2;$$

Then we have, indeed, within the approximation considered:

$$S^*\, S = S\, S^* = 1, \quad S^* = S^{-1}.$$

We assume for S':

$$(10.6)\qquad S' = \frac{i}{h}\, c\, g'\, V^{-1/2} \sum_k \frac{1}{\omega_k^2} \sum_\sigma p_{\sigma,k} \sum_n \tau_n^{(\sigma)}\, e^{-i k x_n},$$

This expression for S' satisfies the condition (10.4) on account of (5.2). It follows by commuting with H^0 on account of the commutation rules (5.4):

$$[H^0, S'] = -\frac{c\,g'}{2} V^{-1/2} \sum_k \sum_\sigma \left(q^*_{\sigma,k} \sum_n \tau_n^{(\sigma)} e^{-ikx_n} + q_{\sigma,k} \sum_n \tau_n^{(\sigma)} e^{+ikx_n} \right),$$

or with (5.2):

$$[H^0, S'] \doteq -H'. \tag{10.7}$$

If we expand the transformed Hamiltonian operator:

$$S^{-1} H S = \left(1 - S' + \frac{1}{2} S'^2 + \ldots\right)\left(H^0 + H'\right)\left(1 + S' + \frac{1}{2} S'^2 + \ldots\right)$$

in powers of g', the linear terms in g' cancel on account of (10.7) and one obtains in second approximation:

$$S^{-1} H S = H^0 + \frac{1}{2} [H', S'] + \ldots. \tag{10.8}$$

The new perturbation term $\frac{1}{2}[H', S']$, appears according to (10.2,.6) as a double sum in n and n'. We separate the terms of the sum $n = n'$ and $n \neq n'$:

$$\frac{1}{2} [H', S'] = H''_{(=)} + H''_{(\neq)}. \tag{10.9}$$

$H''_{(=)}$ is simply the sum of those terms which would also appear, if only one single nucleon were present (self-energy plus terms describing, for instance, scattering processes). These terms of course do not contribute anything to the nuclear forces, at least not in the approximation discussed here. We evaluate therefore only $H''_{(\neq)}$, for which one obtains, on account of $\left[\tau_n^{(\sigma)}, \tau_{n'}^{(\sigma')}\right] = 0$ for $n \neq n'$:

$$H''_{(\neq)} = -c^2 g'^2 \cdot \frac{1}{2} \sum_{n \neq n'} \sum_\sigma \tau_n^{(\sigma)} \tau_{n'}^{(\sigma)} \cdot \frac{1}{V} \sum_k \frac{e^{ik(x_{n'} - x_n)}}{\omega_k^2}.$$

Here again appears the Yukawa potential, as defined by (7.3). We shall write the sum with respect to σ as symbolic scalar product:

$$\sum_\sigma \tau_n^{(\sigma)} \tau_{n'}^{(\sigma)} = \tau_n \cdot \tau_{n'}, \tag{10.10}$$

It follows then:

$$H''_{(\neq)} = -g'^2 \sum_{n < n'} (\tau_n \cdot \tau_{n'})\, U(x_{n'} - x_n). \tag{10.11}$$

We have in (10.10) already discussed the terms $\sigma = 1, 2$ of the matrix $(\tau_n \cdot \tau_{n'})$ [cf. (9.22)]. Here we have an additional term $\tau_n^{(3)} \tau_{n'}^{(3)}$, which according to (9.12) is a diagonal matrix with the elements ± 1. If we denote the four charge states of the particle pair n, n'—as in §9—by a, b, c, d, the following scheme results for the matrix $(\tau_n \cdot \tau_{n'})$:

	a	b	c	d
a	+1	0	0	0
b	0	−1	+2	0
c	0	+2	−1	0
d	0	0	0	+1

The eigen values are: 1, (-1 ± 2), 1. The three eigen values +1 belong to eigen functions that are symmetrical in λ_n, $\lambda_{n'}$ while the eigen function which corresponds to the eigen value −3 is antisymmetrical. If only two proton-neutrons are present, the eigen values of $H''_{(\neq)}$ are:

$$V^s_{12} = -g'^2 \, U(x_1 - x_2), \qquad V^a_{12} = +3 g'^2 \, U(x_1 - x_2), \tag{10.12}$$

where V^s_{12}, V^a_{12} refer to states with symmetrical and antisymmetrical eigen functions in λ_1, λ_2, respectively. In the first case the forces are attractive, in the second repelling. The characteristic difference from the forces discussed in §9 lies in the fact that the three symmetrical states have the same eigen value: the forces between two protons or between two neutrons are the same as those between one proton and one neutron, provided that their eigen function is symmetrical with respect to their charge numbers. In this case one speaks of "charge-independent forces." The symmetrical theory of Kemmer is of special interest in view of the experimental reasons for the charge independence of the nuclear forces (cf. §15).

The charge independence of the symmetrical theory (10.1) is true, not only in the second approximation considered here, but also in any higher approximations (Kemmer). This follows from the fact that the theory is invariant with respect to simultaneous orthogonal transformations of the vectors ψ_1, ψ_2, ψ_3 and $\tau_n^{(1)}$, $\tau_n^{(2)}$, $\tau_n^{(3)}$ (rotations in the space of the isotopic spin).[1] For then the Hamiltonian, once made diagonal with respect to the field quantum numbers, can depend only on the charge numbers by means of the invariant combinations $(\tau_n \cdot \tau_{n'})$, which have eigen values independent of the charge.

§ 11. Charged Particles in an Electromagnetic Field

In classical relativistic mechanics the motion of a particle with the charge e_1 in a given electromagnetic field can be described by a Hamiltonian partial differential equation of the following form:

$$\sum_{\nu=1}^{4} \left(\frac{\partial W}{\partial x_\nu} - \frac{e_1}{c} \Phi_\nu \right)^2 + m^2 c^2 = 0;$$

Here Φ_ν denotes the four-vector potential of the electromagnetic field $F_{\mu\nu}$ taken at the position of the particle:

[1] This holds also for the relations (9.13).

$$F_{\mu\nu} = \frac{\partial \Phi_\nu}{\partial x_\mu} - \frac{\partial \Phi_\mu}{\partial x_\nu}.$$

The wave equation corresponding to this problem is the equation of Schrödinger[1] and Gordon:[2]

$$\left\{ \sum_\nu \left(\frac{\partial}{\partial x_\nu} - \frac{i\, e_1}{h\, c} \Phi_\nu \right)^2 - \left(\frac{m\, c}{h} \right)^2 \right\} \psi = 0.$$

This differential equation, which reduces for the field-free case ($\Phi_\nu = 0$) to (8.1), describes the influence of the electromagnetic field on the charged meson field, discussed in §8. If we consider here the potential components Φ_ν as given space-time functions, it means that we ignore the reaction of the meson field on the electromagnetic field.

Following the notations in §8 we set the mass of the particle $m = \mu h/c$ and the elementary charge $e_1 = \epsilon h$. Then the above wave equation and its conjugate are:

$$(11.1) \quad \left\{ \sum_\nu \left(\frac{\partial}{\partial x_\nu} - \frac{i\, \varepsilon}{c} \Phi_\nu \right)^2 - \mu^2 \right\} \psi = 0, \quad \left\{ \sum_\nu \left(\frac{\partial}{\partial x_\nu} + \frac{i\, \varepsilon}{c} \Phi_\nu \right)^2 - \mu^2 \right\} \psi^* = 0.$$

They can be derived from the following Lagrangian according to (3.3):

$$(11.2) \quad L = -c^2 \left\{ \sum_\nu \left(\frac{\partial \psi^*}{\partial x_\nu} + \frac{i\, \varepsilon}{c} \Phi_\nu\, \psi^* \right) \left(\frac{\partial \psi}{\partial x_\nu} - \frac{i\, \varepsilon}{c} \Phi_\nu\, \psi \right) + \mu^2\, \psi^*\, \psi \right\}.$$

According to (3.13) it follows for the charge and current density of the ψ-field:[3]

$$(11.3) \quad s_\nu = i\, \varepsilon\, c^2 \left(\frac{\partial \psi^*}{\partial x_\nu} \psi - \frac{\partial \psi}{\partial x_\nu} \psi^* \right) - 2\, \varepsilon^2\, c\, \Phi_\nu\, \psi^*\, \psi.$$

The continuity equation $\sum_\nu \partial s_\nu / \partial x_\nu = 0$ is satisfied since L is obviously in-

---

[1] *Ann. d. Phys. 81*, 109, 1926 (§6).

[2] *Z. Phys. 40*, 117, 1926.

[3] Since $\partial\psi/\partial x_\nu$ and $\partial\psi^*/\partial x_\nu$ appear in L only in the combinations:

$$\frac{\partial \psi}{\partial x_\nu} - \frac{i\, \varepsilon}{c} \Phi_\nu\, \psi, \qquad \frac{\partial \psi^*}{\partial x_\nu} + \frac{i\, \varepsilon}{c} \Phi_\nu\, \psi^*$$

it follows from (3.13) that:

$$s_\nu = c \frac{\partial L}{\partial \Phi_\nu}.$$

variant with respect to the gauge transformation (3.10). This is connected with the fact that the theory is also invariant with respect to the more general gauge transformation:

$$(11.4) \qquad \psi \to \psi \cdot e^{\frac{i\varepsilon}{c}\Lambda}, \quad \psi^* \to \psi^* \cdot e^{-\frac{i\varepsilon}{c}\Lambda}, \quad \Phi_\nu \to \Phi_\nu + \frac{\partial \Lambda}{\partial x_\nu},$$

where Λ is any real space-time function. This invariance is necessary since the transformation of the potentials Φ_ν according to (11.4)[1] leaves the field strengths $F_{\mu\nu}$ invariant and since the effects of the field are solely determined by the $F_{\mu\nu}$. One can easily verify[2] that not only the wave equations (11.1) but also the Lagrangian (11.2) and the four-current (11.3) are invariant with respect to the transformation (11.4).

The canonical energy-momentum tensor, constructed according to (3.8), does not yet satisfy this postulate of invariance; furthermore, it is not symmetrical. As mentioned in §2, we shall complete it to a symmetrical and gauge-invariant tensor:

$$(11.5) \qquad T_{\mu\nu} = c^2 \left\{ \left(\frac{\partial \psi^*}{\partial x_\mu} + \frac{i\varepsilon}{c} \Phi_\mu \psi^* \right) \left(\frac{\partial \psi}{\partial x_\nu} - \frac{i\varepsilon}{c} \Phi_\nu \psi \right) + \left(\frac{\partial \psi^*}{\partial x_\nu} + \frac{i\varepsilon}{c} \Phi_\nu \psi^* \right) \left(\frac{\partial \psi}{\partial x_\mu} - \frac{i\varepsilon}{c} \Phi_\mu \psi \right) \right\} + L\, \delta_{\mu\nu}.$$

The divergence of this tensor is, according to the wave equation (11.1), equal to the forces which the electromagnetic field exerts on the mesons.[3]

$$(11.6) \qquad \sum_\mu \frac{\partial T_{\mu\nu}}{\partial x_\mu} = -\frac{1}{c} \sum_\mu s_\mu F_{\mu\nu}.^4$$

---

[1] "Gauge transformation of the second kind" according to Pauli; cf. footnote 1, p. 14.

[2] Notice the operator equation:

$$\left(\frac{\partial}{\partial x_\nu} \mp \frac{i\varepsilon}{c} \left(\Phi_\nu + \frac{\partial \Lambda}{\partial x_\nu} \right) \right) e^{\pm \frac{i\varepsilon}{c}\Lambda} = e^{\pm \frac{i\varepsilon}{c}\Lambda} \left(\frac{\partial}{\partial x_\nu} \mp \frac{i\varepsilon}{c} \Phi_\nu \right).$$

[3] For proving (11.6) it is convenient to use the equation deduced from (11.1):

$$\sum_\mu \frac{\partial}{\partial x_\mu} \left\{ \chi \left(\frac{\partial \psi}{\partial x_\mu} - \frac{i\varepsilon}{c} \Phi_\mu \psi \right) \right\} = \sum_\mu \left(\frac{\partial \chi}{\partial x_\mu} + \frac{i\varepsilon}{c} \Phi_\mu \chi \right) \left(\frac{\partial \psi}{\partial x_\mu} - \frac{i\varepsilon}{c} \Phi_\mu \psi \right) + \mu^2 \chi \psi.$$

[4] In order that this equation may also be valid in the quantized theory and with the sequence

We shall retain for the Hamiltonian the canonical definition (3.6); thus we obtain the canonical field equations in agreement with (11.1). With:

$$\pi = \frac{1}{ic} \frac{\partial L}{\partial \frac{\partial \psi}{\partial x_4}} = \dot{\psi}^* - i\,\varepsilon\,\Phi_0\,\psi^*, \qquad \pi^* = \dot{\psi} + i\,\varepsilon\,\Phi_0\,\psi \tag{11.7}$$

$(\Phi_4 = i\,\Phi_0)$ one finds:

$$H = \pi^*\pi + c^2 \left\{ \sum_{k=1}^{3} \left(\frac{\partial \psi^*}{\partial x_k} + \frac{i\,\varepsilon}{c}\,\Phi_k\,\psi^* \right) \left(\frac{\partial \psi}{\partial x_k} - \frac{i\,\varepsilon}{c}\,\Phi_k \psi \right) + \mu^2 \psi^* \psi \right\} - \Phi_0 \cdot i\,\varepsilon\,(\pi\,\psi - \pi^*\,\psi^*). \tag{11.8}$$

H differs by the last term $\Phi_0\,\rho$ from the energy density $-T_{44}$. One verifies that (11.8) yields the correct field equations. On account of the commutation rules (3.9) there follows:

$$\dot{\psi} = \frac{i}{h} \left[\int dx\, H, \psi \right] = \pi^* - i\,\varepsilon\,\Phi_0\,\psi,$$

in agreement with (11.7). A short computation gives:

$$\dot{\pi}^* = c^2 \left\{ \sum_k \left(\frac{\partial}{\partial x_k} - \frac{i\,\varepsilon}{c}\,\Phi_k \right)^2 - \mu^2 \right\} \psi - i\,\varepsilon\,\Phi_0\,\pi^*.$$

One regains from these equations the wave equation (11.1) by elimination of π^*. If we separate in H the Hamiltonian H^0 of the unperturbed meson field, as it was done in (9.1), we obtain for the perturbation function:

$$H' = H - H^0 = -\,\Phi_0 \cdot i\,\varepsilon\,(\pi\,\psi - \pi^*\,\psi^*) - i\,\varepsilon\,c \sum_{k=1}^{3} \Phi_k \left(\frac{\partial \psi^*}{\partial x_k}\,\psi - \frac{\partial \psi}{\partial x_k}\,\psi^* \right) + \varepsilon^2\,\psi^*\,\psi \sum_{k=1}^{3} \Phi_k^2, \tag{11.9}$$

---

of factors chosen in (11.3) or (8.3) (cf. footnote p. 51), the single terms in $T_{\mu\nu}$ still must be made symmetrical; for instance, in T_{4k},

$$\dot{\psi}^* \left(\frac{\partial \psi}{\partial x_k} - \frac{i\,\varepsilon}{c}\,\Phi_k\,\psi \right)$$

must be replaced by:

$$\frac{1}{2} \left\{ \dot{\psi}^* \left(\frac{\partial \psi}{\partial x_k} - \frac{i\,\varepsilon}{c}\,\Phi_k\,\psi \right) + \left(\frac{\partial \psi}{\partial x_k} - \frac{i\,\varepsilon}{c}\,\Phi_k\,\psi \right) \dot{\psi}^* \right\}.$$

and by comparison with (11.3):

$$H' = \Phi_0 \varrho - \frac{1}{c} \sum_{k=1}^{3} \Phi_k s_k - \varepsilon^2 \psi^* \psi \sum_{k=1}^{3} \Phi_k^2. \tag{11.10}$$

In the quantized theory[1] we use again the matrix representation (5.5, 8.12, 8.13), which makes H^0 diagonal. H' appears then as a bilinear form in the matrices a_k, a_k^*, b_k, b_k^*. This means the perturbation matrix gives rise to transitions between the stationary states of the unperturbed meson field, such that two of the quantum numbers $\overset{+}{N}_k$, $\overset{-}{N}_k$, change by either $+1$ or -1. The total charge of the mesons must be preserved during these transitions, since the electromagnetic field cannot absorb any charges. (The conservation of the total charge $e = \int dx \rho$ follows from the fact that it commutes with H.) Thus the electromagnetic field causes transitions of the following kind:

$$\begin{aligned}
&(\alpha)\ \overset{+}{N}_k \rightarrow \overset{+}{N}_k - 1, \quad \overset{+}{N}_{k'} \rightarrow \overset{+}{N}_{k'} + 1;\\
&(\beta)\ \overset{-}{N}_k \rightarrow \overset{-}{N}_k - 1, \quad \overset{-}{N}_{k'} \rightarrow \overset{-}{N}_{k'} + 1;\\
&(\gamma)\ \overset{+}{N}_k \rightarrow \overset{+}{N}_k + 1, \quad \overset{-}{N}_{k'} \rightarrow \overset{-}{N}_{k'} + 1;\\
&(\delta)\ \overset{+}{N}_k \rightarrow \overset{+}{N}_k - 1, \quad \overset{-}{N}_{k'} \rightarrow \overset{-}{N}_{k'} - 1.
\end{aligned}$$

The processes (α) or (β) represent the scattering of a positive or negative meson by the electromagnetic field; (γ) or (δ) on the other hand describe the pair production or annihilation of a positive and a negative meson. These processes take place as real processes, if they are consistent with energy conservation. They can also appear as virtual transitions in a higher approximation of the perturbation theory.

We treat here as simplest example the scattering of a meson by an electrostatic field:

$$\frac{\partial \Phi_0}{\partial t} = 0, \qquad \Phi_1 = \Phi_2 = \Phi_3 = 0. \tag{11.11}$$

The scalar potential Φ_0 shall have the Fourier expansion:

$$\Phi_0 = \sum_k A_k e^{ikx}, \quad A_{-k} = A_k^*. \tag{11.12}$$

[1] We shall not prove here the Lorentz invariance of the quantization and refer instead to similar considerations in §21 (Electrons in the Electromagnetic Field).

Then it follows from (11.8) with the help of (5.5) by integration over the periodicity volume V:

$$H' = \int_V dx\, H' = -i\varepsilon \sum_k \sum_{k'} A_{k'-k} (p_{k'} q_k - p_k^* q_{k'}^*);$$

splitting off the terms $k = k'$, it follows with (8.11, 8.12):

$$H' = A_0 e + \frac{\varepsilon h}{2} \sum_{k \neq k'} \sum A_{k'-k} \left\{ \frac{\omega_{k'} + \omega_k}{\sqrt{\omega_{k'} \omega_k}} (a_{k'}^* a_k - b_k^* b_{k'}) + \frac{\omega_{k'} - \omega_k}{\sqrt{\omega_{k'} \omega_k}} (a_{k'}^* b_k^* - a_k b_{k'}) \right\}. \tag{11.13}$$

We consider the scattering of a positive meson: its momentum in the initial state shall be hk, in the final state hk'. The corresponding matrix element of H' is according to (8.13) contained in the term $\sim a_{k'}^* a_k$, and has the value, if no other mesons in these states are present, $(\overset{+}{N}_k = 1 \rightarrow 0,\ \overset{+}{N}_{k'} = 0 \rightarrow 1)$:

$$H'_{k'k} = \frac{\varepsilon h}{2} A_{k'-k} \cdot \frac{\omega_{k'} + \omega_k}{\sqrt{\omega_{k'} \omega_k}},$$

Since from the conservation of energy $\omega_{k'} = \omega_k$ (elastic scattering):

$$H'_{k'k} = \varepsilon h A_{k'-k} = \varepsilon h \cdot \frac{1}{V} \int_V dx\, \Phi_0\, e^{i(k-k')x}. \tag{11.14}$$

This result corresponds entirely to the "Born approximation" in the ordinary wave mechanical scattering theory:

$$\left(H'_{FI} = \varepsilon h \int dx\, u_{k'}^* \Phi_0 u_k, \quad \text{where} \quad u_k = V^{-1/2} e^{ikx} \right).$$

If in particular Φ_0 is the Coulomb field of a point charge at rest, the scattering for unrelativistic meson velocities is in agreement with Rutherford's scattering formula. For arbitrary velocities one finds, according to well-known methods, the following cross-section for the scattering in the solid angle $d\Omega$:

$$dQ = d\Omega \cdot \left(1 - \frac{v^2}{c^2}\right) \left(\frac{Z e_1^2}{2 m v^2}\right)^2 \cdot \frac{1}{\sin^4 \frac{\vartheta}{2}} \tag{11.15}$$

(v = meson velocity, ϑ = scattering angle, $e_1 = \epsilon h$ = meson charge, Ze_1 = charge of the scattering nucleus).

The terms $\sim a_{k'}^{*} b_{k}^{*}$ in (11.13) represent the pair production of a positive and a negative meson. The static field Φ_0 of course cannot possibly provide the energy $h(\omega_k + \omega_{k'})$ necessary for the real process. On the other hand a plane light wave alone, even with sufficient frequency, cannot produce a meson pair. For, if one inserts the vector potential of the light wave in relation (11.9), one can easily see that $k' - k$ is determined by the wave-number vector of the light wave. This means the total momentum of the meson pair $(hk') + (-hk)$ is fixed by the momentum of the light quantum. In that case the energy law cannot be fulfilled (on account of $c|k - k'| < \omega_k + \omega_{k'}$). The situation is different if a static field Φ_0 is present beside the high-frequency light wave, for instance, the Coulomb field of a nucleus which can absorb an excess momentum. In this case the second order perturbation theory yields a two-step transition. In this, first a virtual meson pair is produced; in the second process, the momentum of this pair is then changed in such a way that the energy law is satisfied. The cross-section of this process was calculated by Pauli and Weisskopf.[1] We shall not discuss this here. According to this result, one may expect that a light wave of sufficiently high frequency produces charged scalar mesons when passing through matter, provided these particles exist.

We have emphasized above that in these calculations the reaction of the mesons on the electromagnetic field was not included. This is certainly permissible for the Coulomb field of heavy nuclei. In the case of pair production the reaction consists in the disappearance of the incident light quantum. In order to describe this, one should treat the electromagnetic field on the same basis as the meson field, such that H' represents an interaction between the two fields. (Quantum electrodynamics, cf. §§17 ff.) For the case of pair creation, the neglecting of the reaction on the electromagnetic field in the second approximation does not introduce any error. The situation is the same here as in the case of the description of atomic processes induced by the electromagnetic field (absorption and induced emission), where it is sufficient to represent the light wave as a given space-time function. The spontaneous processes on the other hand, as, for instance, the spontaneous annihilation of a meson pair with photon emission, can be described adequately only in quantum electrodynamics.

If the electromagnetic field is such that it cannot produce real pairs on account of the conservation laws, then there exist, nevertheless, virtual pairs. Suppose we start from the state of vacuum (all $\overset{+}{N}_k, \overset{-}{N}_k = 0,\ \Phi_\nu = 0$) and switch on an electrical field adiabatically without adding any real mesons, then the

[1] Cf. reference p. 48.

existence of virtual pairs shows iteslf, for instance, by the fact that the expectation value of the electric charge density $\rho = -i\epsilon(\pi\psi - \pi^*\psi^*)$ may be different from zero, although the space integral naturally vanishes. In this respect the vacuum behaves like a polarizable medium (ρ = density of "free charges"); one speaks of "vacuum polarization." In higher approximations of the perturbation theory the polarizability of the vacuum depends on the field, which has the effect that the superposition principle of electrodynamics no longer holds (non-linearity of the field equations). The energy of the vacuum polarization represents a kind of self-energy of the electromagnetic field, which results from the production of virtual mesons pairs. The calculations based on perturbation theory lead, also here, to divergent k-space integrals, as was found in other self-energy problems (cf. the self-energy of the protons discussed in §7). One obtains meaningful results on the vacuum polarization and similar effects only if the theory is completed by special rules for the removal of these infinities.[1] We shall not enter into this but refer to §21 where analogous phenomena shall be discussed which result from the existence of electrons and positrons (with spin $\frac{1}{2}$).

[1] Cf. Weisskopf, *Kgl. Danske Vid. Selsk, Math.-fys. Medd. XIV*, 6, 1936.

Chapter III

The Vector Meson Field

§ 12. Complex Field in Vacuum[1]

The simplest field next to the scalar field has four components:

$$\psi_1, \psi_2, \psi_3, \psi_4 = i\,\psi_0,$$

which form a four-vector. The classical field with all its components shall again satisfy the Schrödinger-Gordon wave equation (4.16):

$$(\Box - \mu^2)\,\psi_\nu = 0. \tag{12.1}$$

Its solutions shall be restricted by the Lorentz-invariant subsidiary condition:

$$\sum_\nu \frac{\partial \psi_\nu}{\partial x_\nu} = 0 \tag{12.2}$$

so that only three of the wave functions ψ_ν can be considered as independent. From this restriction it follows that the number of stationary states of the quantized field is tripled as compared with the scalar field. We shall see that these three states correspond to the three possible orientations of the meson spin 1. Without the condition (12.2) there would exist one more state with each triplet, which would have to be assigned to a particle of spin 0. This particle would, however, have a negative energy ($-\,h\,\omega_k$ instead of $+\,h\,\omega_k$) if the energy of the particles with spin 1 were chosen positive[2] and such negative

[1] First investigated (without quantization) by Proca, *J. phys. radium* 7, 347, 1936.

[2] With the Lagrangian:

$$L = -\frac{c^2}{2}\left\{\sum_{\mu\nu}\left(\frac{\partial \psi_\nu}{\partial x_\mu}\right)^2 + \mu^2 \sum_\nu \psi_\nu^2\right\},$$

which yields (12.1) without (12.2), one obtains as Hamiltonian:

$$H = \frac{1}{2}\sum_\nu \left\{\pi_\nu^2 + c^2\,(\nabla \psi_\nu)^2 + c^2\,\mu^2\,\psi_\nu^2\right\};$$

the term with $\nu = 4$ in the sum is negative, since $\psi_4 = i\psi_0$ (ψ_0 real).

energies must be excluded.[1] With the definition:

$$f_{\mu\nu} = \frac{\partial \psi_\nu}{\partial x_\mu} - \frac{\partial \psi_\mu}{\partial x_\nu} = - f_{\nu\mu} \tag{12.3}$$

it follows from (12.1, 12.2):

$$\sum_\mu \frac{\partial f_{\mu\nu}}{\partial x_\mu} = \mu^2 \psi_\nu. \tag{12.4}$$

For $\mu = 0$ this corresponds to Maxwell's equations of the electromagnetic field *in vacuo*. We shall see, however, that the case $\mu = 0$ must be treated separately; we shall therefore postpone its discussion until Chapter IV and discuss here only the "meson field" with $\mu \neq 0$.

In view of the applications to the theory of nuclear forces, charged mesons are of greater interest than neutral mesons. We shall, therefore, introduce complex fields from the start. The ψ_ν thus may be complex space-time functions in the classical theory and non-Hermitian operators in the quantized theory. The equations (12.1 and 12.4) shall, of course, be valid also for the ψ_ν^*. Since we use the imaginary time coordinate $x_4 = i\,c\,t$, we must define with $-\psi_4^*$ the complex conjugate or the Hermitian conjugate to ψ_4.

$$\psi_4 = + i\,\psi_0, \quad \psi_4^* = + i\,\psi_0^*;$$

With this definition one can replace in (12.1 to 12.4) ψ everywhere by ψ^* and f by f^*.

We use the following Lagrangian for the canonical formulation of the problem:

$$L = -\frac{1}{2} \sum_{\mu\nu} \left(\frac{\partial \psi_\nu^*}{\partial x_\mu} - \frac{\partial \psi_\mu^*}{\partial x_\nu} \right) \left(\frac{\partial \psi_\nu}{\partial x_\mu} - \frac{\partial \psi_\mu}{\partial x_\nu} \right) - \mu^2 \sum_\nu \psi_\nu^* \psi_\nu. \tag{12.5}$$

The field equations (3.3) can, of course, be applied to all complex field components $\psi = \psi_\nu$ and give in this case:

$$\frac{\partial L}{\partial \psi_\nu^*} - \sum_\mu \frac{\partial}{\partial x_\mu} \frac{\partial L}{\partial \dfrac{\partial \psi_\nu^*}{\partial x_\mu}} = - \mu^2 \psi_\nu + \sum_\mu \frac{\partial}{\partial x_\mu} \left(\frac{\partial \psi_\nu}{\partial x_\mu} - \frac{\partial \psi_\mu}{\partial x_\nu} \right) = 0; \tag{12.6}$$

[1] Such charged particles of negative energy would carry out transitions into states of higher momentum under emission of light. We shall discuss later why negative energy states are permissible for the Dirac electron, but not in our case (§§20 to 22).

Differentiating with respect to x_ν and summing over ν, we obtain:

$$\mu^2 \sum_\nu \frac{\partial \psi_\nu}{\partial x_\nu} = \sum_{\mu\nu} \frac{\partial^2 f_{\mu\nu}}{\partial x_\mu \partial x_\nu};$$

Since the double sum on the right hand side vanishes on account of $f_{\mu\nu} = -f_{\nu\mu}$, there follows subsidiary condition (12.6) on account of $\mu \neq 0$. With this the equations (12.6) are identical with the wave equations (12.1). The same is true if one interchanges the ψ_ν with the ψ_ν^*. We have thus shown that the Lagrangian (12.5) yields the desired field equations and it is seen that for this result the assumption $\mu \neq 0$ is essential.

In the expression (3.13) for the electric charge and current density one must add the contributions of the four field components:

$$s_\nu = -i\varepsilon \sum_\mu \left(\frac{\partial L}{\partial \frac{\partial \psi_\mu}{\partial x_\nu}} \psi_\mu - \frac{\partial L}{\partial \frac{\partial \psi_\mu^*}{\partial x_\nu}} \psi_\mu^* \right) = i\varepsilon \sum_\mu (f_{\nu\mu}^* \psi_\mu - f_{\nu\mu} \psi_\mu^*). \tag{12.7}$$

One verifies the continuity equation $\sum_\nu \partial s_\nu / \partial x_\nu = 0$ with the help of (12.3, 12.4). We shall discuss the energy-momentum tensor later.

The field variables, which are canonical conjugates to ψ_ν, ψ_ν^*, are:

$$\begin{cases} \pi_\nu = \dfrac{1}{ic} \dfrac{\partial L}{\partial \frac{\partial \psi_\nu}{\partial x_4}} = \dfrac{1}{ic} \left(\dfrac{\partial \psi_4^*}{\partial x_\nu} - \dfrac{\partial \psi_\nu^*}{\partial x_4} \right) = \dfrac{1}{ic} f_{\nu 4}^*, \\ \pi_\nu^* = \dfrac{1}{ic} \dfrac{\partial L}{\partial \frac{\partial \psi_\nu^*}{\partial x_4}} = \dfrac{1}{ic} \left(\dfrac{\partial \psi_4}{\partial x_\nu} - \dfrac{\partial \psi_\nu}{\partial x_4} \right) = \dfrac{1}{ic} f_{\nu 4}. \end{cases} \tag{12.8}$$

Here we have the special case of π_4 and π_4^* vanishing identically.[1]

$$\pi_4 = 0, \quad \pi_4^* = 0. \tag{12.9}$$

In spite of this, we shall form the Hamiltonian in the usual way:

$$H = \sum_\nu (\pi_\nu \dot{\psi}_\nu + \pi_\nu^* \dot{\psi}_\nu^*) - L,$$

[1] L is independent of $\dot{\psi}_4$ and $\dot{\psi}_4^*$. We have drawn attention to this case in §1.

where we put, according to (12.8) ($j = 1, 2, 3$):

$$(12.10) \qquad \dot{\psi}_j = i\,c\,\frac{\partial \psi_j}{\partial x_4} = c^2 \pi_j^* + i\,c\,\frac{\partial \psi_4}{\partial x_j}, \qquad \dot{\psi}_j^* = c^2 \pi_j + i\,c\,\frac{\partial \psi_4^*}{\partial x_j}.$$

If we further add to $\mathbf{H}$ the space divergence:

$$-i\,c \sum_j \frac{\partial}{\partial x_j} (\pi_j \psi_4 + \pi_j^* \psi_4^*)$$

this term does not change the integral Hamiltonian function $H = \int dx\, \mathbf{H}$. We obtain then:

$$(12.11) \qquad \mathbf{H} = c^2 \sum_j \pi_j^* \pi_j + \frac{1}{2} \sum_{ij} \left(\frac{\partial \psi_i^*}{\partial x_j} - \frac{\partial \psi_j^*}{\partial x_i} \right) \left(\frac{\partial \psi_i}{\partial x_j} - \frac{\partial \psi_j}{\partial x_i} \right) + \mu^2 \sum_\nu \psi_\nu^* \psi_\nu - i\,c \left\{ \psi_4 \sum_j \frac{\partial \pi_j}{\partial x_j} + \psi_4^* \sum_j \frac{\partial \pi_j^*}{\partial x_j} \right\}.$$

The indices i, j are running here only from 1 to 3.[1]

In writing down the commutation relations one must take care that the commutators $[\pi_4, \psi_4]$ and $[\pi_4^*, \psi_4^*]$ vanish identically according to (12.9). The functions with the index 4 cannot be considered as true canonical variables. We have, however, the option to eliminate these field components altogether, owing to the fact that only three of the four functions ψ_ν are independent of each other. If one writes thus in (12.11), in accordance with (12.6 and 12.8):

$$(12.12) \qquad \psi_4 = \frac{1}{\mu^2} \sum_j \frac{\partial}{\partial x_j} \left(\frac{\partial \psi_4}{\partial x_j} - \frac{\partial \psi_j}{\partial x_4} \right) = \frac{i\,c}{\mu^2} \sum_j \frac{\partial \pi_j^*}{\partial x_j}, \qquad \psi_4^* = \frac{i\,c}{\mu^2} \sum_j \frac{\partial \pi_j}{\partial x_j},$$

it follows:

$$(12.13) \qquad \mathbf{H} = c^2 \sum_j \pi_j^* \pi_j + \frac{c^2}{\mu^2} \sum_i \frac{\partial \pi_i^*}{\partial x_i} \sum_j \frac{\partial \pi_j}{\partial x_j} + \mu^2 \sum_j \psi_j^* \psi_j + \frac{1}{2} \sum_{ij} \left(\frac{\partial \psi_i^*}{\partial x_j} - \frac{\partial \psi_j^*}{\partial x_i} \right) \left(\frac{\partial \psi_i}{\partial x_j} - \frac{\partial \psi_j}{\partial x_i} \right).$$

In this last expression only the components of the space vectors $\psi = (\psi_1, \psi_2, \psi_3)$ and $\pi = (\pi_1, \pi_2, \pi_3)$, together with their complex conjugates ψ^* and π^*, appear.

[1] Since the identically vanishing function π_4 does not appear in $\mathbf{H}$, $\mathbf{H}$ does not yield the correct canonical field equation for $\dot{\psi}_4$; but one can succeed by adding formally to $\mathbf{H}$ certain terms which contain π_4. We shall not discuss this, since the following elimination of ψ_4, ψ_4^* will lead to the same result.

In the notation of vector analysis the equations (12.12, 12.13) are:

$$\psi_4 = \frac{i\,c}{\mu^2}\,\nabla\cdot\pi^*, \qquad \psi_4^* = \frac{i\,c}{\mu^2}\,\nabla\cdot\pi; \tag{12.14}$$

$$H = c^2\;\pi^*\cdot\pi\; + \frac{c^2}{\mu^2}\;\nabla\cdot\pi^*\;\;\nabla\cdot\pi$$
$$+ \mu^2\;\psi^*\cdot\psi\; + \nabla\times\psi^*\cdot\nabla\times\psi\,. \tag{12.15}$$

It is to be noticed that H depends now on the space derivatives of the π_j, π_j^* [cf. (1.10)]. It is positive-definite, being a sum of only positive terms of the form $\varphi^*\,\varphi$.

According to the canonical formalism we quantize the theory by postulating:

$$[\pi_j\,(x), \psi_j\,(x')] = [\pi_j^*\,(x), \psi_j^*\,(x')] = \frac{h}{i}\,\delta\,(x - x') \quad (j = 1,\,2,\,3), \tag{12.16}$$

while all other pairs of functions with the indices 1, 2, 3 shall commute. One finds then as canonical field equations:

$$\dot\psi_j\,(x) = \frac{i}{h}\int dx'\;[H\,(x'), \psi_j\,(x)]$$
$$= \int dx'\left\{c^2\,\pi_j^*\,(x')\cdot\delta(x' - x) + \frac{c^2}{\mu^2}\;\;\nabla'\cdot\pi^*\,(x')\cdot\frac{\partial}{\partial x_j'}\,\delta(x' - x)\right\},$$

and after partial integration:

$$\dot\psi = c^2\,\pi^* - \frac{c^2}{\mu^2}\,\nabla\,(\nabla\cdot\pi^*)\,; \tag{12.17}$$

correspondingly:

$$\dot\pi^* = -\,\mu^2\,\psi - \nabla\times(\nabla\times\psi). \tag{12.18}$$

By eliminating π^*, respectively ψ from (12.17, 12.18) it follows:

$$\frac{1}{c^2}\,\ddot\psi = -\,\mu^2\,\psi - \nabla\times(\nabla\times\psi) + \nabla(\nabla\cdot\psi) = -\,\mu^2\,\pi^* + \nabla^2\,\pi^*. \tag{12.19}$$

$$\frac{1}{c^2}\,\ddot\pi^* = -\,\mu^2\,\pi^* - \nabla\times(\nabla\times\pi^*) + \nabla\,(\nabla\cdot\pi^*) = -\,\mu^2\,\psi + \nabla^2\,\psi,$$

For the function ψ_4, defined by (12.14), it follows at the same time:

$$\frac{1}{c^2}\,\ddot\psi_4 = -\,\mu^2\,\psi_4 + \nabla^2\,\psi_4, \tag{12.20}$$

and with (12.18):

$$\frac{1}{ic}\dot{\psi}_4 = \frac{1}{\mu^2}\nabla\cdot\dot{\pi}^* = -\nabla\cdot\psi. \tag{12.21}$$

Since the equations (12.19, 12.20, 12.21) agree formally with (12.1, 12.2), we have shown that the Hamiltonian (12.15) together with the commutation relations (12.16) represents the properties of the ψ-field correctly. It also leads to the correct relationship between the fields $\dot{\psi}_\nu$ and π_j, for it follows from (12.17) with the help of (12.14):

$$\pi^* = \frac{1}{c^2}\dot{\psi} + \frac{1}{ic}\nabla\psi_4,$$

which is in agreement with (12.8). It is obvious also that all equations are valid which one obtains from the ones above by interchanging the starred quantities with those without stars. It is to be emphasized that this method of eliminating ψ_4 and $\dot{\psi}_4^*$ rests again essentially on the assumption $\mu \neq 0$.

In order to verify the relativistic invariance of the quantization method, we introduce again time-dependent operators

$$\psi_\nu(x,t) = e^{\frac{it}{h}H}\psi_\nu(x)\,e^{-\frac{it}{h}H}, \qquad \psi_\nu^*(x,t) = e^{\frac{it}{h}H}\psi_\nu^*(x)\,e^{-\frac{it}{h}H},$$

where $H = \int dx\,\mathsf{H}$ is given by (12.15). We consider first the field components $\nu \neq 4$. On account of (12.19 and 4.17) the formulas (4.28,.21) can be applied. Since according to (12.17) $\dot{\psi}$ is connected only with π^*, and $\dot{\psi}^*$ only with π, it follows:

$$[\psi_\nu(x,t), \psi_{\nu'}(x',t')] = [\psi_\nu^*(x,t), \psi_{\nu'}^*(x',t')] = 0 \tag{12.22}$$

(the $d_{\sigma\sigma'}^{(1)} = 0$ in this case). Furthermore:

$$[\psi_\nu(x,t), \psi_{\nu'}^*(x',t')] = \frac{h}{i}d_{\nu\nu'}D(x-x', t-t'), \tag{12.23}$$

where the $d_{\nu\nu'}$ are defined by the equations:

$$\overset{(1)}{\psi_\nu}(x) = \dot{\psi}_\nu(x) = \sum_{\nu'} d_{\nu\nu'}\pi_{\nu'}^*(x) + \ldots$$

The comparison with (12.17) yields:

$$d_{\nu\nu'} = c^2\left\{\delta_{\nu\nu'} - \frac{1}{\mu^2}\frac{\partial^2}{\partial x_\nu \partial x_{\nu'}}\right\}. \tag{12.24}$$

So far these formulas refer only to the components ν, $\nu' = 1, 2, 3$; the considerations of §4 are in this case not valid for the components ψ_4, ψ_4^*, since according to (12.14, 12.16) they do not commute with the ψ_j^*, or ψ_j, in contradiction to the canonical commutation rules of §4 [cf. (4.14)]. Nevertheless the commutation rules (12.22, 12.23, 12.24) should also be valid for $\nu = 4$, or $\nu' = 4$, if they shall be relativistically invariant; for only in that case do the right and the left hand sides transform in the same way, namely, like the product of two 4-vectors or a tensor of 2nd rank. In order to prove the invariance of the quantized theory, we must still show that the formulas (12.22, 12.23, 12.24) are compatible with the equations (12.14 to 12.21), for ν or $\nu' = 4$. We restrict our discussion to that of (12.23) [in connection with (12.24)] since, then, it becomes trivial for (12.22).

If we differentiate—as in §4 [cf. (4.13)]—(12.23) n times with respect to t and n' times with respect to t', and afterwards put: $t = t' = 0$, it follows with the help of the definition (4.7):

$$\left[\overset{(n)}{\psi_\nu}(x), \overset{(n')}{\psi^*_{\nu'}}(x')\right] = \frac{h}{i}(-1)^{n'}\left\{\frac{\partial^{n+n'}}{\partial t^{n+n'}} d_{\nu\nu'} D(x - x', t)\right\}_{t=0}; \tag{12.25}$$

All these equations together are equivalent to (12.23). Their right hand sides can be evaluated with the help of (4.30, 4.31 and 4.33):

$$\frac{\partial^2 D(x,t)}{\partial t^2} = c^2(\nabla^2 - \mu^2) D(x,t), \quad D(x, 0) = 0, \quad \left(\frac{\partial D(x,t)}{\partial t}\right)_{t=0} = \delta(x). \tag{12.26}$$

In particular according to (12.24):

$$d_{44} = c^2 + \frac{1}{\mu^2}\frac{\partial^2}{\partial t^2}, \quad \text{also} \quad d_{44} D(x,t) = \frac{c^2}{\mu^2}\nabla^2 D(x,t)$$

With $n = n' = 0$ (12.25) yields ($j, j' = 1, 2, 3$):

$$[\psi_j(x), \psi^*_{j'}(x')] = [\psi_4(x), \psi^*_4(x')] = 0,$$

$$[\psi_4(x), \psi^*_j(x')] = [\psi_j(x), \psi^*_4(x')] = \frac{hc}{\mu^2}\frac{\partial}{\partial x_j}\delta(x - x'),$$

in agreement with (12.14, 12.16). Furthermore with $n = 1$, $n' = 0$:

$$[\dot\psi_j(x), \psi^*_{j'}(x')] = \frac{h}{i}c^2\left\{\delta_{jj'} - \frac{1}{\mu^2}\frac{\partial^2}{\partial x_j \partial x_{j'}}\right\}\delta(x - x'),$$

$$[\dot\psi_4(x), \psi^*_4(x')] = \frac{h}{i}\frac{c^2}{\mu^2}\nabla^2\delta(x - x'),$$

$$[\dot\psi_4(x), \psi^*_j(x')] = [\dot\psi_j(x), \psi^*_4(x')] = 0;$$

If one expresses here the $\dot{\psi}_j$ and $\dot{\psi}_4$ according to (12.17 and 12.21) by the π_j^* and ψ_j, one recognizes again the agreement with (12.14, 12.16). For $n = n' = 1$ one need not calculate the right hand sides of (12.25) again, since they follow according to (12.26) from the values calculated for $n = n' = 0$ by applying the operator $- c^2 (\nabla^2 - \mu^2)$. Here too we have agreement with (12.14 to 12.21). Higher n values do not give any new equations on account of (12.19, 12.20) and (12.26). This completes the proof of the invariant commutation rules (12.23, 12.24).

For the representation of the Hamiltonian in momentum space we can use the Fourier series (5.5), by identifying ψ, ψ^*, π, π^* with the vector fields entering into (12.15). Here q_k, q_k^*, p_k, p_k^* represent, of course, 3-component vectors. Their components satisfy according to (12.16) the following equations:

$$[p_{k,j}, q_{k',j'}] = [p^*_{k,j}, q^*_{k',j'}] = \frac{h}{i}\, \delta_{kk'}\, \delta_{jj'}. \tag{12.27}$$

With (12.15) one obtains:[1]

$$H = \int_V dx\, \mathsf{H} = \sum_k \left\{ c^2 (p_k^* \cdot p_k) + \frac{c^2}{\mu^2} (k \cdot p_k^*)(k \cdot p_k) + \mu^2 (q_k^* \cdot q_k) + ([k \times q_k^*] \cdot [k \times q_k]) \right\}. \tag{12.28}$$

For the following it is convenient to decompose the individual amplitude vectors (for instance q_k) into longitudinal and transverse components. Let $\mathrm{e}_k^{(1)}$, $\mathrm{e}_k^{(2)}$, $\mathrm{e}_k^{(3)}$ be an orthogonal system of unit vectors:

$$\mathrm{e}_k^{(r)} \cdot \mathrm{e}_k^{(s)} = \delta_{rs}, \tag{12.29}$$

oriented in such a way that $\mathrm{e}_k^{(1)}$ is parallel to the momentum vector k:

$$\begin{cases} \mathrm{e}_k^{(1)} \cdot k = |k|, & \mathrm{e}_k^{(2)} \cdot k = \mathrm{e}_k^{(3)} \cdot k = 0, \\ \mathrm{e}_k^{(1)} \times k = 0, & \mathrm{e}_k^{(2)} \times k = -|k|\, \mathrm{e}_k^{(3)}, \quad \mathrm{e}_k^{(3)} \times k = |k|\, \mathrm{e}_k^{(2)}. \end{cases} \tag{12.30}$$

We denote the components of the amplitude vectors in the directions $\mathrm{e}_k^{(r)}$ by upper indices r $(= 1, 2, 3)$:

$$\begin{cases} q_k = \sum_r \mathrm{e}_k^{(r)} q_k^{(r)}, & q_k^* = \sum_r \mathrm{e}_k^{(r)} q_k^{(r)*}, \\ p_k = \sum_r \mathrm{e}_k^{(r)} p_k^{(r)}, & p_k^* = \sum_r \mathrm{e}_k^{(r)} p_k^{(r)*}; \end{cases} \tag{12.31}$$

[1] $a \cdot b$ signifies the scalar, $a \times b$ the vector product of the 3-component vectors a and b.

According to (12.27, 12.29) one has the commutation rules:

$$[p_k^{(r)}, q_k^{(s)}] = [p_k^{(r)*}, q_k^{(s)*}] = \frac{h}{i}\,\delta_{rs}, \tag{12.32}$$

while all other pairs of components commute. By substituting (12.31) into (12.28), we can separate H into contributions from the longitudinal and the transverse components.

$$\begin{cases} H = H^{\text{long}} + H^{\text{tr}}, \\ H^{\text{long}} = \sum_k \left\{ \dfrac{\omega_k^2}{\mu^2}\, p_k^{(1)*}\, p_k^{(1)} + \mu^2\, q_k^{(1)*}\, q_k^{(1)} \right\}, \\ H^{\text{tr}} = \sum_k \sum_{r=2,3} \left\{ c^2\, p_k^{(r)*}\, p_k^{(r)} + \dfrac{\omega_k^2}{c^2}\, q_k^{(r)*}\, q_k^{(r)} \right\}; \end{cases} \tag{12.33}$$

here ω_k is again defined by (6.15). In (12.33) each individual term k, r of the sum can be represented by a diagonal matrix, similarly as in §8. We write [as in (8.12)]:

$$\begin{cases} q_k^{(1)} = \dfrac{1}{\mu}\sqrt{\dfrac{h\,\omega_k}{2}}\,(a_k^{(1)} + b_k^{(1)*}), & q_k^{(1)*} = \dfrac{1}{\mu}\sqrt{\dfrac{h\,\omega_k}{2}}\,(a_k^{(1)*} + b_k^{(1)}), \\ p_k^{(1)} = \mu\sqrt{\dfrac{h}{2\,\omega_k}}\, i\,(a_k^{(1)*} - b_k^{(1)}), & p_k^{(1)*} = \mu\sqrt{\dfrac{h}{2\,\omega_k}}\, i\,(-a_k^{(1)} + b_k^{(1)*}); \end{cases} \tag{12.34}$$

$$\begin{cases} q_k^{(r)} = c\sqrt{\dfrac{h}{2\,\omega_k}}\,(a_k^{(r)} + b_k^{(r)*}), & q_k^{(r)*} = c\sqrt{\dfrac{h}{2\,\omega_k}}\,(a_k^{(r)*} + b_k^{(r)}), \\ p_k^{(r)} = \dfrac{1}{c}\sqrt{\dfrac{h\,\omega_k}{2}}\, i\,(a_k^{(r)*} - b_k^{(r)}), & p_k^{(r)*} = \dfrac{1}{c}\sqrt{\dfrac{h\,\omega_k}{2}}\, i\,(-a_k^{(r)} + b_k^{(r)*}) \\ & \text{für } r = 2, 3. \end{cases} \tag{12.35}$$

Here the a and b are matrices of the type (6.16) or (8.13) with respect to integers $\overset{+}{N}{}_k^{(r)}$ and $\overset{-}{N}{}_k^{(r)}$, with the commutators:

$$[a_k^{(r)}, a_k^{(r)*}] = [b_k^{(r)}, b_k^{(r)*}] = 1, \tag{12.36}$$

while all other commutators vanish. According to (8.14):

$$(a_k^{(r)*}\, a_k^{(r)})_N = \overset{+}{N}{}_k^{(r)}, \quad (b_k^{(r)*}\, b_k^{(r)})_N = \overset{-}{N}{}_k^{(r)}. \tag{12.37}$$

With (12.34, 12.35) the commutations rules (12.32) are satisfied and one obtains for H (12.33) the diagonal matrix:

$$H = \sum_k \sum_{r=1}^{3} h\,\omega_k\,\frac{1}{2}\,\{a_k^{(r)*}\, a_k^{(r)} + a_k^{(r)}\, a_k^{(r)*} + b_k^{(r)*}\, b_k^{(r)} + b_k^{(r)}\, b_k^{(r)*}\}. \tag{12.38}$$

Similarly as in (8.17) one obtains for the eigen values of H:

(12.39) $$H_N = \sum_k h\omega_k \sum_{r=1}^{3} (\overset{+}{N}{}_k^{(r)} + \overset{-}{N}{}_k^{(r)}) + 6\,H_0.$$

For the field momentum one obtains from (2.10) and (3.7):[1]

(12.40) $$G_j = \int dx\, \boldsymbol{G}_j = -\int dx \sum_\nu \left(\pi_\nu \frac{\partial \psi_\nu}{\partial x_j} + \frac{\partial \psi_\nu^*}{\partial x_j} \pi_\nu^* \right),$$

or with (5.5) and (12.9):

(12.41) $$G = -i \sum_k k \left\{ (p_k \cdot q_k) - (q_k^* \cdot p_k^*) \right\}.$$

The matrix representation by means of (12.31, 12.34, 12.35) yields, as in (8.10, 8.16), the diagonal matrix:

(12.42) $$G = \sum_k \sum_{r=1}^{3} h\, k \left\{ a_k^{(r)*}\, a_k^{(r)} - b_k^{(r)}\, b_k^{(r)*} \right\},$$

with the eigen values:

(12.43) $$G_N = \sum_k h\, k \sum_{r=1}^{3} (\overset{+}{N}{}_k^{(r)} - \overset{-}{N}{}_k^{(r)}).$$

For the total charge of the field one obtains from (3.11) or from (12.7, 12.8):

(12.44) $$\begin{aligned} e = \int dx\, \varrho &= -i\,\varepsilon \int dx \left\{ \pi \cdot \psi - \pi^* \cdot \psi^* \right\} \\ &= -i\,\varepsilon \sum_k \left\{ p_k \cdot q_k - p_k^* \cdot q_k^* \right\} \\ &= \frac{h\,\varepsilon}{2} \sum_k \sum_{r=1}^{3} \left\{ a_k^{(r)*}\, a_k^{(r)} + a_k^{(r)}\, a_k^{(r)*} - b_k^{(r)*}\, b_k^{(r)} - b_k^{(r)}\, b_k^{(r)*} \right\}; \end{aligned}$$

(12.45) $$e_N = h\,\varepsilon \sum_k \sum_{r=1}^{3} (\overset{+}{N}{}_k^{(r)} - \overset{-}{N}{}_k^{(r)}).$$

From the eigen values (12.39, 12.43, 12.45) it follows that a stationary state characterized by the quantum numbers $\overset{+}{N}{}_k^{(r)}, \overset{-}{N}{}_k^{(r)}$ contains $(\overset{+}{N}{}_k^{(1)} + \overset{+}{N}{}_k^{(2)} + \overset{+}{N}{}_k^{(3)})$ positive mesons of momentum hk and $(\overset{-}{N}{}_k^{(1)} + \overset{-}{N}{}_k^{(2)} + \overset{-}{N}{}_k^{(3)})$ negative mesons

[1] The generalization of this formula for the case in question (several complex field components) is trivial. About the momentum density see below.

of momentum $-hk$. The number of mesons of given charge and given momentum is the sum of three independent integers. We have therefore a three-fold degeneracy and the "vector meson" has still another determining factor, which can assume three values. This must be a spin coordinate. That this is so can be verified easily by considering the three states of a (positive or negative) meson at rest. Since for $k = 0$ the orientation of the axes $\mathfrak{e}_0^{(1)}, \mathfrak{e}_0^{(2)}, \mathfrak{e}_0^{(3)}$ is arbitrary and the distinction between longitudinal and transverse polarization is impossible, one has in this case three stationary states, which can be transferred into each other by rotations of the coordinate system. This characterizes particles of spin 1. The following discussion of the angular momentum will confirm this conclusion.

A state of the total system characterized by the set of quantum numbers $\overset{+}{N}_k^{(r)}, \overset{+}{N}_k^{(r)}$ has the statistical weight 1, corresponding to Bose-Einstein statistics.

We shall discuss now the form of the energy-momentum tensor. The canonical tensor, here called T^0, which was constructed according to the rule (2.8), can be written according to (12.5 and .3):

$$(12.46) \qquad T^0_{\mu\nu} = -\sum_\varrho \left(\frac{\partial L}{\partial \dfrac{\partial \psi_\varrho}{\partial x_\mu}} \frac{\partial \psi_\varrho}{\partial x_\nu} + \frac{\partial \psi_\varrho^*}{\partial x_\nu} \frac{\partial L}{\partial \dfrac{\partial \psi_\varrho^*}{\partial x_\mu}} \right) + L\,\delta_{\mu\nu}$$

$$= \sum_\varrho \left(f^*_{\mu\varrho} \frac{\partial \psi_\varrho}{\partial x_\nu} + \frac{\partial \psi_\varrho^*}{\partial x_\nu} f_{\mu\varrho} \right) + L\,\delta_{\mu\nu};$$

It satisfies the conservation law (2.7):

$$(12.47) \qquad \sum_\mu \frac{\partial T^0_{\mu\nu}}{\partial x_\mu} = 0.$$

Since it is evidently not symmetrical, we complete it to a symmetrical tensor by adding a further tensor:

$$(12.48) \qquad T'_{\mu\nu} = -\sum_\varrho \frac{\partial}{\partial x_\varrho} (f^*_{\mu\varrho} \psi_\nu + \psi^*_\nu f_{\mu\varrho}),$$

The divergence of $T'_{\mu\nu}$ vanishes identically because it has the form $\sum_{\mu\rho} a_{\mu\rho}$, with $a_{\mu\rho} = -a_{\rho\mu}$.

$$(12.49) \qquad \sum_\mu \frac{\partial T'_{\mu\nu}}{\partial x_\mu} = 0.$$

For the sum of the tensors (12.46 and 12.48) one obtains with (12.3, 12.4):

$$T_{\mu\nu} = T^0_{\mu\nu} + T'_{\mu\nu} \tag{12.50}$$
$$= \sum_\varrho (f^*_{\mu\varrho} f_{\nu\varrho} + f^*_{\nu\varrho} f_{\mu\varrho}) + \mu^2 (\psi^*_\mu \psi_\nu + \psi^*_\nu \psi_\mu) + L . \delta_{\mu\nu}.$$

Now the symmetry condition:

$$T_{\mu\nu} = T_{\nu\mu} \tag{12.51}$$

is satisfied, and at the same time we have, according to (12.47, 12.49):

$$\sum_\mu \frac{\partial T_{\mu\nu}}{\partial x_\mu} = 0. \tag{12.52}$$

The sequence of factors in (12.50) is chosen in such a way that in the quantized theory the operators which belong to real field quantities are Hermitian. According to (2.6) we can now interpret $-T_{44}$ as energy density and T_{4j}/ic as components of the momentum density. On account of (12.5,.8 and .14), $-T_{44}$ equals the Hamiltonian (12.15), and hence is also positive-definite. On the other hand the momentum density T_{4j} is different from the canonical density T^0_{4j}, used in (12.40). The difference $T_{4j} - T^0_{4j}$ is, however, according to (12.48), a space divergence $\left(\sum_{j'} \partial\varphi_{jj'}/\partial x_{j'},\ j' \neq 4\right)$, which does not contribute anything to the space integral:

$$\int dx\, T_{4j} = \int dx\, T^0_{4j}; \tag{12.53}$$

This is why we could calculate the total momentum G (12.40 to 12.43) without error from the canonical tensor. However, for the computation of the angular momentum:

$$M = \int dx\, x \times G \tag{12.54}$$

it is necessary to use the complete expression for the momentum density:

$$G = G^0 + G', \quad G^0_j = \frac{1}{ic} T^0_{4j}, \quad G'_j = \frac{1}{ic} T'_{4j}. \tag{12.55}$$

We decompose the angular momentum in a corresponding way:

$$M = M^0 + M', \quad M^0 = \int dx\, x \times G^0, \quad M' = \int dx\, x \times G'. \tag{12.56}$$

One obtains for the components of the two partial moments, with the help of (12.46, .48 and .8):

$$M^0_{jj'} = -\int dx \sum_i \left\{ x_j \left(\pi_i \frac{\partial \psi_i}{\partial x_{j'}} + \frac{\partial \psi_i^*}{\partial x_{j'}} \pi_i^* \right) - x_{j'} \left(\pi_i \frac{\partial \psi_i}{\partial x_j} + \frac{\partial \psi_i^*}{\partial x_j} \pi_i^* \right) \right\}, \tag{12.57}$$

$$\begin{aligned} M'_{jj'} &= \int dx \sum_i \left\{ x_j \frac{\partial}{\partial x_i} (\pi_i \psi_{j'} + \psi_{j'}^* \pi_i^*) - x_{j'} \frac{\partial}{\partial x_i} (\pi_i \psi_j + \psi_j^* \pi_i^*) \right\} \\ &= -\int dx \left\{ (\pi_j \psi_{j'} + \psi_{j'}^* \pi_j^*) - (\pi_{j'} \psi_j + \psi_j^* \pi_{j'}^*) \right\} \end{aligned} \tag{12.58}$$

(the last expression for M' is obtained by partial integration). Substituting the Fourier series (5.5) and using vector notation, we have:

$$M^0 = -\frac{i}{V} \int_V dx \sum_{kk'} e^{i(k-k')x} \{ x \times k \ (p_{k'} \cdot q_k) - x \times k' \ (q_{k'}^* \cdot p_k^*) \}, \tag{12.59}$$

$$M' = -\sum_k \{ p_k \times q_k - q_k^* \times p_k^* \}. \tag{12.60}$$

The separation of M into M^0 and M' has a physical meaning since—as we maintain—M^0 represents the orbital angular momentum and M' the spin angular momentum of the meson, provided that the velocities of all mesons present are of a non-relativistic order of magnitude. One can show that the expectation value of M^0 is independent of the state of polarization of the meson, i.e., it does not change when a meson with constant momentum is transferred into any other state of polarization; hence M^0 can in no way be connected with the spin of the meson. We shall here forego the proof, which requires some computations. We shall show instead that the spin angular momentum, which is of main interest to us, is represented by M'. Since M' is, according to (12.60), of the form $\sum_k M'_{(k)}$, it is sufficient to consider each term of the sum separately. We examine especially the term $M'_{(0)}$ $(k = 0)$, which means that we shall compute the spin of a meson at rest. We have already remarked that the orientation of the coordinate axes $e_k^{(1)}$, $e_k^{(2)}$, $e_k^{(3)}$ is entirely arbitrary in the rest system. The distinction between longitudinal and transverse components is no longer possible and the three axes are equivalent. This is also expressed by the fact that the transformation formulae (12.34, 12.35) will become identical for $r = 1, 2, 3$ (with $k = 0$, $\omega_0 = c\mu$ it follows $\sqrt{\omega_0}/\mu = c/\sqrt{\omega_0}$). Suppressing the index $k = 0$, we write for the component of $M'_{(0)}$ in the direction of the arbitrarily chosen $\mathfrak{e}^{(1)}$-axis:

$$M'_1 = -p^{(2)} q^{(3)} + p^{(3)} q^{(2)} + q^{(2)*} p^{(3)*} - q^{(3)*} p^{(2)*},$$

or with (12.35):

$$M'_1 = h i \{-a^{(2)*} a^{(3)} + a^{(3)*} a^{(2)} - b^{(2)*} b^{(3)} + b^{(3)*} b^{(2)}\}.$$

The contributions of the positive and negative mesons (a- and b-terms) are separated here, and in view of their symmetry it is sufficient to consider only the positive mesons. We put thus $\overline{N}^{(r)} = 0$, so that the b-terms are zero:

$$M'_1 = h i \{-a^{(2)*} a^{(3)} + a^{(3)*} a^{(2)}\}. \tag{12.61}$$

Instead of the operators $a^{(2)}$, $a^{(2)*}$, $a^{(3)}$, $a^{(3)*}$, we now introduce four new operators a_+, a_+^*, a_-, a_-^* by the transformation:[1]

$$\begin{cases} a_+ = \dfrac{1}{\sqrt{2}} (a^{(2)} - i\, a^{(3)}), & a_+^* = \dfrac{1}{\sqrt{2}} (a^{(2)*} + i\, a^{(3)*}), \\ a_- = \dfrac{1}{\sqrt{2}} (a^{(2)} + i\, a^{(3)}), & a_-^* = \dfrac{1}{\sqrt{2}} (a^{(2)*} - i\, a^{(3)*}). \end{cases} \tag{12.62}$$

From (12.36) it follows:

$$[a_+, a_+^*] = [a_-, a_-^*] = 1,$$

whereas all other pairs formed from the four new operators commute. Thus, since the transformation (12.62) leaves the commutation rules invariant, the new operators can again be represented by matrices of the type (6.16):

$$(a_+)_{N'_+ N'_-, N''_+ N''_-} = (a_+^*)_{N''_+ N''_-, N'_+ N'_-} = \sqrt{N''_+} \cdot \delta_{N'_+, N''_+ - 1} \cdot \delta_{N'_-, N''_-},$$

and in the same way for a_- a_-^*. Here $a_+^* a_+$ and $a_-^* a_-$ become now diagonal matrices with the (non-negative) eigen values N_+, N_-:

$$(a_+^* a_+)_{N_+ N_-} = N_+, \quad (a_-^* a_-)_{N_+ N_-} = N_-.$$

[1] In the unquantized theory, a_+ and a_- represent the amplitudes of circular vibrations.

According to (12.62) we have:

$$\overset{*}{a}_+ a_+ + \overset{*}{a}_- a_- = a^{(2)*} a^{(2)} + a^{(3)*} a^{(3)}$$

or

$$N_+ + N_- = \overset{+}{N}{}^{(2)} + \overset{+}{N}{}^{(3)}.$$

On the other hand (12.62) yields:

$$\overset{*}{a}_+ a_+ - \overset{*}{a}_- a_- = i\,(- a^{(2)*} a^{(3)} + a^{(3)*} a^{(2)}) = \frac{M'_1}{h};$$

M'_1 is thus diagonal in the scheme of the quantum numbers N_+, N_- and has the eigen values:

$$(M'_1)_{N_+ N_-} = h\,(N_+ - N_-). \tag{12.63}$$

This result may be interpreted as follows: the component of the meson spin angular momentum in any arbitrary fixed direction in space can assume the values $\pm h$ and 0. If $\overset{+}{N}{}^{(1)} + \overset{+}{N}{}^{(2)} + \overset{+}{N}{}^{(3)} = \overset{+}{N}{}^{(1)}_1 + N_+ + N_-$ is the total number of mesons, $\overset{+}{N}{}^{(1)}_1$ have the spin components 0, N_+ have the components $+h$ and N_- have the components $-h$. The three values for the components correspond to the three possibilities of orientation of a meson with total spin h. For the square of the spin angular momentum $M'_{(0)}$, i.e., of the term $k = 0$ of the sum in (12.60), one obtains on account of (12.61):

$$M'^2 = \sum_j M'^2_j = -\frac{h^2}{2} \sum_{r \neq s} (a^{(r)*} a^{(s)} - a^{(s)*} a^{(r)})^2$$

$$= h^2 \sum_{r \neq s} \{- (a^{(r)*} a^{(s)})^2 + a^{(r)*} a^{(r)} \cdot a^{(s)} a^{(s)*}\}.$$

If we consider here only such states in which only a single positive meson at rest exists $\left(\sum_r \overset{+}{N}{}^{(r)} = 1\right)$, the operators $(a^{(s)})^2$ which diminish the meson number $\overset{+}{N}{}^{(s)}$ by 2 yield zero when applied to the Schrödinger function of such states. The remaining terms in M'^2 are diagonal in the scheme of the quantum numbers $\overset{+}{N}{}^{(r)}$ and yield:

$$M'^2 = h^2 \sum_{r \neq s} \overset{+}{N}{}^{(r)} (\overset{+}{N}{}^{(s)} + 1);$$

here the terms $\overset{+}{N}^{(r)} \cdot \overset{+}{N}^{(s)}$ vanish also, since one of the two factors is always zero, and it remains:

$$M'^2 = h^2 \cdot 2 \sum_r \overset{+}{N}^{(r)} = h^2 \cdot 2.$$

This corresponds to the known value $h^2 s\,(s+1)$ of the square of the angular momentum with $s = 1$, and it confirms again the spin value 1 of the mesons discussed here.

Besides the mechanical spin-angular momentum, the charged vector meson has also a magnetic moment. In order to calculate this, too, we shall investigate in the following section the effect of an external electromagnetic field on the meson field.

§ 13. Vector Mesons in the Electromagnetic Field[1]

We consider the Maxwell field:

$$F_{\mu\nu} = \frac{\partial \Phi_\nu}{\partial x_\mu} - \frac{\partial \Phi_\mu}{\partial x_\nu} \tag{13.1}$$

as a given external field, acting on the charge of the mesons. In order to obtain in this case the Lagrangian of the vector meson field, we notice, that in the case of a scalar field the Lagrangian (11.2) results from that for $\Phi = 0$ (8.2), if one replaces

$$\frac{\partial \psi}{\partial x_\nu} \text{ by } \left(\frac{\partial \psi}{\partial x_\nu} - \frac{i\,\varepsilon}{c}\,\Phi_\nu\,\psi\right) \text{ and } \frac{\partial \psi^*}{\partial x_\nu} \text{ by } \left(\frac{\partial \psi^*}{\partial x_\nu} + \frac{i\,\varepsilon}{c}\,\Phi_\nu\,\psi^*\right)$$

where $\epsilon\,h$ signifies the elementary charge e_1. We use here the same rule with regard to all components of the fields ψ_ν and ψ_ν^*. For the operators which take the place of $\partial/\partial x_\nu$ we use the abbreviations:

$$\partial_\nu \equiv \frac{\partial}{\partial x_\nu} - \frac{i\,\varepsilon}{c}\,\Phi_\nu, \quad \partial_\nu^* \equiv \frac{\partial}{\partial x_\nu} + \frac{i\,\varepsilon}{c}\,\Phi_\nu. \tag{13.2}$$

Thus one obtains from (12.5) the Lagrangian:

$$\begin{aligned} L &= -\frac{1}{2}\sum_{\mu\nu}(\partial_\mu^*\,\psi_\nu^* - \partial_\nu^*\,\psi_\mu^*)\,(\partial_\mu\,\psi_\nu - \partial_\nu\,\psi_\mu) - \mu^2\sum_\nu \psi_\nu^*\,\psi_\nu \\ &= -\frac{1}{2}\sum_{\mu\nu} f_{\mu\nu}^*\,f_{\mu\nu} - \mu^2 \sum_\nu \psi_\nu^*\,\psi_\nu, \end{aligned} \tag{13.3}$$

[1] See footnote 1, p. 75.

with the new definitions:

$$f_{\mu\nu} \equiv \partial_\mu \psi_\nu - \partial_\nu \psi_\mu, \quad f^*_{\mu\nu} \equiv \partial^*_\mu \psi^*_\nu - \partial^*_\nu \psi^*_\mu \tag{13.4}$$

One obtains now, as can easily be seen with the help of (13.2), the field equations:

$$\frac{\partial L}{\partial \psi^*_\nu} - \sum_\mu \frac{\partial}{\partial x_\mu} \frac{\partial L}{\partial \frac{\partial \psi^*_\nu}{\partial x_\mu}} = -\mu^2 \psi_\nu + \sum_\mu \partial_\mu f_{\mu\nu} = 0. \tag{13.5}$$

From this it follows for $\mu \neq 0$:

$$\sum_\nu \partial_\nu \psi_\nu = \frac{1}{\mu^2} \sum_{\nu\mu} \partial_\nu \partial_\mu f_{\mu\nu} = \frac{1}{\mu^2} \cdot \frac{1}{2} \sum_{\nu\mu} [\partial_\nu, \partial_\mu] f_{\mu\nu};$$

This expression is in general not zero, since according to (13.2,.1):

$$[\partial_\nu, \partial_\mu] = \frac{i\varepsilon}{c} \left(\frac{\partial \Phi_\nu}{\partial x_\mu} - \frac{\partial \Phi_\mu}{\partial x_\nu} \right) = \frac{i\varepsilon}{c} F_{\mu\nu}; \tag{13.6}$$

consequently:

$$\sum_\nu \partial_\nu \psi_\nu = \frac{i\varepsilon}{2c\mu^2} \sum_{\mu\nu} F_{\mu\nu} f_{\mu\nu},$$

$$\sum_\nu \frac{\partial \psi_\nu}{\partial x_\nu} = \frac{i\varepsilon}{c} \left\{ \sum_\nu \Phi_\nu \psi_\nu + \frac{1}{2\mu^2} \sum_{\mu\nu} F_{\mu\nu} f_{\mu\nu} \right\}.$$

For the current density vector and the energy-momentum tensor, we can simply take over the formulas (12.7 and 12.50), where $f_{\mu\nu}$ and $f^*_{\mu\nu}$ must be interpreted according to the new definitions (13.4).[1] We obtain now:

$$\sum_\mu \frac{\partial T_{\mu\nu}}{\partial x_\mu} = -\frac{1}{c} \sum_\mu s_\mu F_{\mu\nu}.^{2}$$

[1] For the current density:

$$s_\nu = c \cdot \partial L / \partial \Phi_\nu$$

Cf. footnote 3, p. 67.

[2] Since:

$$\frac{\partial}{\partial x_\mu} (\chi \varphi) = (\partial^*_\mu \chi) \varphi + \chi (\partial_\mu \varphi),$$

it follows from (12.50) with the help of (13.5):

$$\sum_\mu \frac{\partial T_{\mu\nu}}{\partial x_\mu} = \sum_{\mu\varrho} f^*_{\mu\varrho} \left(-\frac{1}{2} \partial_\nu f_{\mu\varrho} + \partial_\mu f_{\nu\varrho} \right) + \mu^2 \left(\sum_\mu \partial^*_\mu \psi^*_\mu \right) \psi_\nu + \text{ conj.}$$

Instead of this one can also write:

L, s_ν and $T_{\mu\nu}$ are evidently invariant under the gauge transformation (11.4).

Just as in the scalar theory (§11), the energy density $-T_{44}$ as Hamiltonian does not yield the correct field equations. For that reason we shall go back to the canonical definition of the Hamiltonian:

$$H = \sum_\nu (\pi_\nu \dot{\psi}_\nu + \pi_\nu^* \dot{\psi}_\nu^*) - L,$$

where:

$$\pi_\nu = \frac{\partial L}{\partial \dot{\psi}_\nu} = \frac{1}{i c} f_{\nu 4}^*, \quad \pi_\nu^* = \frac{1}{i c} f_{\nu 4}, \tag{13.7}$$

$$\begin{cases} \dot{\psi}_\nu = i c \left(\partial_4 + \frac{i \varepsilon}{c} \Phi_4\right) \psi_\nu = c^2 \pi_\nu^* + i c\, \partial_\nu \psi_4 - i \varepsilon \Phi_0 \psi_\nu, \\ \dot{\psi}_\nu^* = i c \left(\partial_4^* - \frac{i \varepsilon}{c} \Phi_4\right) \psi_\nu^* = c^2 \pi_\nu + i c\, \partial_\nu^* \psi_4^* + i \varepsilon \Phi_0 \psi_\nu^* \end{cases} \tag{13.8}$$

($\Phi_4 = i \Phi_0$). If one adds to H the divergence:

$$-i c \sum_j \frac{\partial}{\partial x_j} (\pi_j \psi_4 + \psi_4^* \pi_j^*) = -i c \sum_j \{\partial_j^* \pi_j \cdot \psi_4 + \pi_j \cdot \partial_j \psi_4 + \partial_j \pi_j^* \cdot \psi_4^* + \pi_j^* \cdot \partial_j^* \psi_4^*\},$$

and if one eliminates ψ_4, ψ_4^* with the help of the wave equation (13.5) using the fact that $\pi_4 = \pi_4^* = 0$

$$\psi_4 = \frac{1}{\mu^2} \sum_i \partial_j f_{j4} = \frac{i c}{\mu^2} \sum_j \partial_j \pi_j^*, \quad \psi_4^* = \frac{i c}{\mu^2} \sum_j \partial_j^* \pi_j, \tag{13.9}$$

one obtains finally the Hamiltonian:

$$\sum_\mu \frac{\partial T_{\mu\nu}}{\partial x_\mu} = -\frac{1}{2} \sum_{\mu\varrho} f_{\mu\varrho}^* \left(\partial_\nu f_{\mu\varrho} + \partial_\varrho f_{\nu\mu} + \partial_\mu f_{\varrho\nu} + \frac{i\varepsilon}{c} F_{\mu\varrho} \psi_\nu\right) + \text{conj.}$$

$$= -\frac{1}{2} \sum_{\mu\varrho} f_{\mu\varrho}^* \left([\partial_\nu, \partial_\mu] \psi_\varrho + [\partial_\varrho, \partial_\nu] \psi_\mu + [\partial_\mu, \partial_\varrho] \psi_\nu + \frac{i\varepsilon}{c} F_{\mu\varrho} \psi_\nu\right) + \text{conj.},$$

or with (13.6):

$$\sum_\mu \frac{\partial T_{\mu\nu}}{\partial x_\mu} = -\frac{i\varepsilon}{c} \sum_\mu F_{\mu\nu} \sum_\varrho f_{\mu\varrho}^* \psi_\varrho + \text{conj.}$$

But this is according to (12.7) the stipulated conservation equation. In the quantized theory the sequence of factors in $T_{\mu\nu}$ must be arranged suitably, similarly as in the scalar theory (cf. footnote 4, p. 68).

(13.10) $$H = c^2 \sum_j \pi_j^* \pi_j + \frac{c^2}{\mu^2} \sum_i \partial_i \pi_i^* \cdot \sum_j \partial_j^* \pi_j + \mu^2 \sum_j \psi_j^* \psi_j + \frac{1}{2} \sum_{ij} f_{ij}^* f_{ij} - \Phi_0 \cdot i \varepsilon \sum_j (\pi_j \psi_j - \pi_j^* \psi_j^*).$$

H differs by the last term ($\Phi_0 \cdot \rho$) from the energy density $- T_{44}$ (12.50). With the commutation rules (12.16) the following field equations follow from (13.10):[1]

$$\dot{\psi}_j = \frac{i}{h} [H, \psi_j] = c^2 \pi_j^* - \frac{c^2}{\mu^2} \partial_j \sum_i \partial_i \pi_i^* - i \varepsilon \Phi_0 \psi_j,$$

which agrees with (13.8, 13.9). Furthermore:

$$\dot{\pi}_j^* = \frac{i}{h} [H, \pi_j^*] = - \mu^2 \psi_j + \sum_i \partial_i f_{ij} - i \varepsilon \Phi_0 \pi_j^*$$

or:

$$i c \cdot \partial_4 \pi_j^* \equiv - \partial_4 f_{4j} = - \mu^2 \psi_j + \sum_i \partial_i f_{ij},$$

according to the field equations (13.5). Combined with the equations (13.9) defining ψ_4, ψ_4^* the Hamiltonian (13.10) is thus equivalent to the Lagrangian (13.3).

With this formulation for the interaction of an electromagnetic field with the meson field we shall now determine the magnetic moment of the meson. We introduce for this purpose a weak, stationary, homogeneous magnetic field $\mathfrak{H}$, represented as the curl of a vector potential (3-vector) Φ:

$$\mathfrak{H} = \nabla \times \Phi \qquad \Phi = \tfrac{1}{2} (\mathfrak{H} \times x) \qquad (\mathfrak{H} = \text{const.})^2$$

We expand the Hamiltonian (13.10), with the help of (13.4, .2) in powers of $\mathfrak{H}$ or Φ, and we shall be interested only in those terms which are linear in Φ. In vector notations these can be represented as follows:

$$H_1 = - 2 \sum_j \Gamma_j \Phi_j = - 2 (\Gamma \cdot \Phi),$$

where:

$$\Gamma = - \frac{i \varepsilon}{2 c} \left\{ \frac{c^2}{\mu^2} [\pi (\nabla \cdot \pi^*) - (\nabla \cdot \pi) \pi^*] + [\psi^* \times (\nabla \times \psi) + (\nabla \times \psi^*) \times \psi] \right\}$$

[1] Notice that $\overset{*}{\partial}_j$ changes into $-\partial_j$ by partial integration.

[2] At the same time an electric field $\mathfrak{E} = - \nabla \Phi_0$ (for instance a central field) can also be considered.

Since $\Gamma \cdot (\mathfrak{H} \times x) = \mathfrak{H} \cdot (x \times \Gamma)$, it follows that:

$$H_1 = -\mathfrak{H} \cdot \mathfrak{M}, \quad \text{where} \quad \mathfrak{M} = x \times \Gamma .$$

According to this, $\mathfrak{M}$ is the density of the magnetic moment in the limit $\mathfrak{H} \to 0$. We compare this density with that of the mechanical moment $\mathbf{M} = (x \times \mathbf{G})$, restricting ourselves to the case of slow mesons only. In the limit $\mathfrak{H} = 0$, one obtains for $\mathbf{G}$ ($\mathbf{G}_j = T_{4j}/i\,c$) from (12.50, 12.8, 12.12):

$$\mathbf{G} = \psi^* (\nabla \cdot \pi^*) + (\nabla \cdot \pi) \psi + (\nabla \times \psi^*) \times \pi^* - \pi \times (\nabla \times \psi).$$

We should compare Γ with $\mathbf{G}$. To accomplish this, we again introduce the Fourier series (5.5) and the matrix representation (12.34, 12.35), setting in the last $\omega_k = c\,\mu$ in the sense of a non-relativistic approximation. The result of the computation is written down, jointly for Γ and $\mathbf{G}$, in a form where the upper sign refers to Γ, the lower to $\mathbf{G}$:

$$\left.\begin{matrix} \dfrac{2\mu}{\varepsilon}\,\Gamma = \\ G = \end{matrix}\right\} \frac{h}{2}\frac{1}{V}\sum_{kk'} e^{i(k-k')x} \sum_{rr'} \Big\{ e_{k'}^{(r')} (k \cdot e_k^{(r)}) (a_{k'}^{(r')*} \mp b_{k'}^{(r')}) (a_k^{(r)} - b_k^{(r)*})$$

$$+ e_k^{(r)} (k' \cdot e_{k'}^{(r')}) (a_{k'}^{(r')*} - b_{k'}^{(r')}) (a_k^{(r)} \mp b_k^{(r)*})$$

$$+ e_{k'}^{(r')} \times (k \times e_k^{(r)}) (a_{k'}^{(r')*} \pm b_{k'}^{(r')}) (a_k^{(r)} + b_k^{(r)*})$$

$$+ e_k^{(r)} \times (k' \times e_{k'}^{(r')}) (a_{k'}^{(r')*} + b_{k'}^{(r')}) (a_k^{(r)} \pm b_k^{(r)*}) \Big\}.$$

If one now calculates expectation values for states in which only positive (or only negative) mesons are present, those terms which contain factors b, b^* (or a, a^*) will vanish, and one obtains:

$$\Gamma = \frac{\varepsilon}{2\mu} G, \qquad \mathfrak{M} = \frac{\varepsilon}{2\mu} M \qquad \text{for positive mesons}$$

$$\Gamma = \frac{-\varepsilon}{2\mu} G, \qquad \mathfrak{M} = \frac{-\varepsilon}{2\mu} M \qquad \text{for negative mesons.}$$

Thus there exists according to this theory exact proportionality between the magnetic and mechanical moment for slow mesons of definite charge; the ratio of the two moments is $\pm\,\epsilon/2\mu = \pm\,e_1/2mc$ ($\epsilon h = e_1$ = elementary charge, $\mu h/c = m$ = mass of meson). This is not only true for the orbital moment, but also for the spin moment, which means that the magentic spin moment with regard to its numerical value equals a "meson-magneton"

$e_1h/2mc$. In fact the eigen values of a component of the magnetic spin moment of either positive or negative mesons at rest are for the limit $\Phi_\nu \to 0$ according to (12.63):

$$\left(\int dx\, \mathfrak{M}_1'\right)_{N_+ N_-} = \frac{\pm \varepsilon}{2\mu} (M_1')_{N_+ N_-} = \frac{\pm e_1 h}{2mc} (N_+ - N_-).$$

It should, however, be noted that the equation (13.3) for the coupling of the mesons with the Maxwell field is not unambiguous. One can add to L a term:

$$\gamma \frac{i\varepsilon}{2c} \sum_{\mu\nu} F_{\mu\nu} (\psi_\mu^* \psi_\nu - \psi_\nu^* \psi_\mu)$$

where γ is an arbitrary, real constant.[1] This changes the magnetic spin moment by the factor $1 - \gamma$; i.e., in the system at rest its value will be $(1 - \gamma)$ times as large as that of a meson-magneton.

The effects of weak electromagnetic fields on the mesons of spin 1 can be found with methods of the perturbation theory as in the case of scalar mesons. As an example we shall discuss here again the scattering of positive mesons by an electrostatic field Φ_0 (11.11, 11.12). According to (13.10) the perturbation function for this problem is:

$$H' = \int dx\, \Phi_0\, \varrho = -i\varepsilon \int dx\, \Phi_0 \sum_j (\pi_j \psi_j - \pi_j^* \psi_j^*),$$

or with the Fourier decompositions according to (11.12 and 5.5):

$$H' = -i\varepsilon \sum_{kk'} A_{k'-k} \{(p_{k'} \cdot q_k) - (p_k^* \cdot q_{k'}^*)\}.$$

After separation in longitudinal and transverse components according to (12.31), we introduce the matrix representations (12.34, 12.35), omitting all terms which contain b or b^*, thus excluding negative mesons. In this way we obtain for H' an expression of the following form:

$$H' = \varepsilon h \sum_{kk'} A_{k'-k} \sum_{rr'} (\mathrm{e}_k^{(r)} \cdot \mathrm{e}_{k'}^{(r')})\, \zeta_{k'k}^{(r',r)}\, a_{k'}^{(r')*} a_k^{(r)}; \tag{13.11}$$

[1] Pauli, Solvay report, 1939, *unpublished; Rev. Modern Phys. 13*, 203, 1941; Corben and Schwinger, *Phys. Rev. 58*, 953, 1940.

On account of the conservation of energy the coefficients ζ are needed only for $\omega_{k'} = \omega_k$ $(k' \neq k)$.

$$(13.12) \begin{cases} \zeta_{k'k}^{(r,r)} = \zeta_{k'k}^{(2,3)} = \zeta_{k'k}^{(3,2)} = 1, \\ \zeta_{k'k}^{(1,2)} = \zeta_{k'k}^{(2,1)} = \zeta_{k'k}^{(1,3)} = \zeta_{k'k}^{(3,1)} = \dfrac{(\mu c)^2 + \omega_k^2}{2\mu c \ \omega_k} \equiv \zeta_k \quad (\omega_{k'} = \omega_k). \end{cases}$$

For the initial state, we consider one meson with a definite momentum k and a definite polarization r ($\overset{+}{N}{}_k^{(r)} = 1$, all other $N = 0$). In this case the probability for a transition into a definite final state k', r' is determined by the square of the matrix element:

$$H_{k'k}^{\prime(r',r)} = \varepsilon h A_{k'-k} (\mathbf{e}_k^{(r)} \cdot \mathbf{e}_{k'}^{(r')}) \zeta_{k'k}^{(r',r)}.$$

Thus the transition probability does depend not only on the momenta k, k' but also on the polarizations r, r'. If we disregard the polarization of the scattered radiation, i.e., if we are interested only in its total intensity in the direction k', we must compute $\sum_{r'} |H_{k'k}^{\prime(r',r)}|^2$. Considering first the case that the incident meson is polarized longitudinally ($r = 1$), we have with the help of (13.12):

$$\sum_{r'} \{(\mathbf{e}_k^{(1)} \cdot \mathbf{e}_{k'}^{(r')}) \zeta_{k'k}^{(r',1)}\}^2 = 1 + (\zeta_k^2 - 1) \sin^2 \vartheta,$$

where ϑ is the scattering angle: $\cos\vartheta = \mathbf{e}_k^{(1)} \cdot \mathbf{e}_{k'}^{(1)}$. If one considers further, that [with $\omega_k^2 = c^2 (\mu^2 + k^2)$]:

$$\zeta_k^2 - 1 = \frac{c^2 k^4}{4\mu^2 \omega_k^2},$$

it follows:

$$(13.13) \qquad \sum_{r'} |H_{k'k}^{\prime(r',1)}|^2 = (\varepsilon h)^2 |A_{k'-k}|^2 \left(1 + \frac{c^2 k^4}{4\mu^2 \omega_k^2} \sin^2\vartheta\right).$$

For transversely polarized mesons, on the other hand ($r = 2$), one has according to (13.12):

$$\sum_{r'} \{(\mathbf{e}_k^{(2)} \cdot \mathbf{e}_{k'}^{(r')}) \zeta_{k'k}^{(r',2)}\}^2 = 1 + (\zeta_k^2 - 1) \cos^2\varphi,$$

where $\cos\varphi = \mathbf{e}_k^{(2)} \cdot \mathbf{e}_{k'}^{(1)}$, i.e., φ is the angle which the direction of scattering k' forms with the polarization vector $\mathbf{e}_k^{(2)}$ of the incident meson. For this case one finds:

$$(13.14) \qquad \sum_{r'} \left|H'^{(r',2)}_{k'k}\right|^2 = (\varepsilon h)^2 \left|A_{k'-k}\right|^2 \left(1 + \frac{c^2 k^4}{4 \mu^2 \omega_k^2} \cos^2 \varphi\right).$$

The case of general polarization of the initial state can also be treated with the help of (13.11), but we shall not enter into this here.[1] For non-relativistic meson velocities ($|k| \ll \mu$), where $c^2k^4/\mu^2 \omega_k^2$ can be neglected against 1, one arrives in all cases at the Rutherford formula for the scattering by the Coulomb field of a point charge. At higher velocities we have, compared with the scattering of the scalar meson (11.14, 11.15), an additional scattering, which is determined by the ϑ-, or by the φ-dependent terms in (12.13, 12.14) and can be considered as an effect of the Coulomb field on the moving magnetic spin moment.[2] At very high energies ($|k| \gg \mu$, $\omega_k \gg c\,\mu$) this additional scattering prevails, i.e., the meson with the spin 1 will be much more strongly scattered than that with the spin 0. The formal reason for this lies in the fact that the coefficients ζ_k (13.12), which determine the transition probabilities between longitudinal and transverse meson states increase without limit with increasing energy $h\,\omega_k$. Also for other effects of the electromagnetic fields, for instance the production of meson pairs by light quanta in the presence of a Coulomb field (cf. §11),[3] the cross-sections at high energy are much larger than the corresponding cross-sections in the scalar theory.

§ 14. Nuclear Interactions[4]

Beside the electromagnetic proton-meson interactions, which do not change the total charge of the meson field, there can also exist "nuclear" interactions between mesons and nucleons such as were studied in §9 for the scalar case. They cause the emission or absorption of charged mesons accompanied by a corresponding change of the charge of the nucleon such that the total charge is

[1] Cf. Laporte, *Phys. Rev.* *54*, 905, 1938, where the scattering of the vector mesons in the Coulomb field is discussed on the basis of the non-quantized theory.

[2] This effect shows itself also for the binding of a meson in an attractive force field. In case of the Coulomb field of a point charge ($\Phi_0 \sim r^{-1}$), for certain values of the angular momentum quantum numbers there exist no wave functions which satisfy the usual regularity conditions. Cf. Corben and Schwinger, *loc. cit.*

[3] Cf. Booth and Wilson, *Proc. Roy. Soc. London* *175*, 483, 1940 (p. 513); Kobayasi and Utiyama, *Proc. Phys.-Math. Soc. Japan* *22*, 882, 1940; Christy and Kusaka, *Phys. Rev.* *59*, 405, 1941.

[4] Yukawa, Sakata, and Taketani, *Proc. Phys.-Math. Soc. Japan* *20*, 319, 1938; Kemmer, *Proc. Roy. Soc. London*, *166*, 127, 1938; Fröhlich, Heitler, and Kemmer, *ibid.* *166*, 154, 1938; Stueckelberg, *Helv. Phys Acta* *11*, 299, 1938; Bhabha, *Proc. Roy. Soc. London* *166*, 501, 1938; Heitler, *ibid.* *166*, 529, 1938.

conserved. For reasons of simplicity we shall again omit the electrical forces from our discussion of this interaction in the case of vector mesons. If the coupling term in the Lagrangian is assumed to be linear in the field functions ψ_ν, ψ^*_ν, the simplest interaction term will be of the form:

$$L' = \sum_\nu (\eta_\nu \psi_\nu + \overset{*}{\eta}_\nu \overset{*}{\psi}_\nu),$$

where η_ν represents a 4-vector, which has to be constructed from the wave functions of the nucleons. The components η_1, η_2, η_3 are proportional to the current density or the velocity of the nucleons, and they will vanish in the limiting case of infinitely heavy nucleons:

$$L' = \eta_4 \psi_4 + \overset{*}{\eta}_4 \overset{*}{\psi}_4.$$

With $L = L^0 + L'$, where L^0 is given by (12.5), the field equation (12.6) will be changed for $\nu = 4$ in the following way:

$$\sum_\mu \frac{\partial f_{\mu 4}}{\partial x_\mu} - \mu^2 \psi_4 + \overset{*}{\eta}_4 = 0. \tag{14.1}$$

In constructing the Hamiltonian, formulas (12.8, 12.9, 12.10) remain valid, so that only the term $-L'$ must be added to (12.11). If one eliminates ψ_4 and ψ_4^* with (14.1), in consideration of $\pi_4 = \pi_4^* = 0$, one finds for H, apart from an additive space divergence:

$$H = H^0 + H' + H'',$$

where H^0 is given by (12.13 or 12.15) and where:

$$H' = \frac{c}{\mu^2}\{\eta_0 (\nabla \cdot \pi^*) + \overset{*}{\eta}_0 (\nabla \cdot \pi)\}, \qquad H'' = \frac{1}{\mu^2} \overset{*}{\eta}_0 \eta_0 \tag{14.2}$$

($\eta_4 = i\,\eta_0$, $\overset{*}{\eta}_4 = i\,\overset{*}{\eta}_0$). For η_0, $\overset{*}{\eta}_0$ we assume a point interaction of the type (7.3) or (9.2):

$$\eta_0 = \sum_n g_n\, \delta(x - x_n), \qquad \overset{*}{\eta}_0 = \sum_n g_n^*\, \delta(x - x_n), \tag{14.3}$$

where the g_n, g_n^* denote matrices with respect to the proton charge numbers λ_n, which must be selected in such a way that the conservation of charge is maintained. This condition is satisfied, if we again identify g_n and g_n^* in (14.3) with the "isotopic spin-matrices" (9.11, 9.15); the proof is the same as in §9.[1] With

[1] In fact, the equations (9.3 to 9.15) hold unchanged [in (9.4) $\pi\psi$ must, according to (12.44), be interpreted as the scalar product of the vectors π and ψ].

(5.5), (12.30, 12.31) we have:

$$(14.4)\quad H' = \int_V dx\, \boldsymbol{H}' = \frac{i\,c}{\mu^2}\, V^{-1/2} \sum_k |k| \left\{ p_k^{(1)*} \sum_n g_n\, e^{i k x_n} - p_k^{(1)} \sum_n g_n^* \, e^{-i k x_n} \right\}.$$

Since H'' does not contain the field variables at all, and since H' depends only on the $p_k^{(1)}$, $p_k^{(1)*}$, the expression (14.2) couples only the longitudinal mesons with the stationary nucleons. For the transverse mesons the formulas of §12 remain strictly valid.

In order to calculate the nuclear forces with perturbation theory, based on the interaction term (14.2), we shall proceed as in §10: we transform the Hamiltonian with a unitary matrix S, which is expanded in rising powers of the coupling parameter [cf. (10.3, 10.4, 10.5)]:

$$S = 1 + S' + S'' + \ldots, \qquad S'^* = -S', \qquad S'' = S''^* = \frac{1}{2}\, S'^2.$$

where:

$$(14.5)\qquad S' = \frac{c}{h}\, V^{-1/2} \sum_k \frac{|k|}{\omega_k^2} \left\{ q_k^{(1)} \sum_n g_n\, e^{i k x_n} - q_k^{(1)*} \sum_n g_n^*\, e^{-i k x_n} \right\};$$

with the help of this equation and with (12.33) it follows analogously to (10.7):

$$[H^0, S'] = [H^{\mathrm{long}}, S'] = -H',$$

so that the transformed Hamiltonian up to terms of the second order in the coupling parameter can be represented as follows [cf. (10.8)]:

$$S^{-1} H\, S = S^* (H^0 + H' + H'')\, S = H^0 + \frac{1}{2}\, [H', S'] + H'' + \ldots.$$

According to (14.2, 14.3):

$$(14.6)\qquad H'' = \int_V dx\, \boldsymbol{H}'' = \frac{1}{\mu^2} \sum_{n n'} g_n^*\, g_{n'}\, \delta\,(x_{n'} - x_n);$$

If we split H'' like $[H', S']$ into terms $n = n'$ and $n \neq n'$:

$$(14.7)\qquad \frac{1}{2}\, [H', S'] + H'' = H''_{(=)} + H''_{(\neq)},$$

we get with (14.4, 14.5, 14.6):

$$H''_{(\neq)} = \frac{1}{\mu^2} \sum_{n \neq n'} g_n^*\, g_{n'} \left\{ -\frac{1}{V} \sum_k \frac{c^2\, k^2}{\omega_k^2}\, e^{ik(x_{n'} - x_n)} + \delta\,(x_{n'} - x_n) \right\}.$$

Considering further that:

$$\delta\,(x_{n'} - x_n) = \frac{1}{V} \sum_k e^{ik(x_{n'} - x_n)}$$

and that $\omega_k^2 = c^2(\mu^2 + k^2)$, one obtains the simple result:

$$H''_{(\neq)} = c^2 \sum_{n \neq n'} g_n^* g_{n'} \frac{1}{V} \sum_k \frac{e^{ik(x_{n'} - x_n)}}{\omega_k^2} = \sum_{n \neq n'} g_n^* g_{n'} U(x_{n'} - x_n), \tag{14.8}$$

where U again signifies the Yukawa potential function, defined by (7.13). The comparison with the terms $n \neq n'$ in (9.19) shows that the longitudinal vector mesons produce the same nuclear forces, except for the sign, as the scalar mesons. The signs are everywhere opposite. Everything which was said in §9 about the nuclear forces transmitted by charged mesons (pp. 60 ff.) is also true here—apart from the change of sign.

In computing the terms $H''_{(=)}$ in (14.7) it must be considered that g_n^* and g_n do not commute. One has according to (9.13, 9.15):

$$[g_n^*, g_n] = |g|^2 \tau_n^{(3)}.$$

Omitting self-energy terms, which do not depend on the field variables, we find:

$$H''_{(=)} = -\frac{i c^2}{2 h \mu^2} |g|^2 \cdot \frac{1}{V} \sum_{k k'} |k| \, |k'| \left(\frac{q_k^{(1)} p_{k'}^{(1)}}{\omega_k^2} - \frac{q_{k'}^{(1)*} p_k^{(1)*}}{\omega_{k'}^2} \right) \sum_n \tau_n^{(3)} e^{i(k-k')x_n}.$$

This perturbation function describes, according to the matrix representation (12.34), a scattering of longitudinal mesons by nuclear particles, and in addition double emission and absorption processes. It follows for the matrix element for the elastic scattering of a longitudinal meson on a neutron ($\tau^{(3)} = 1$) or proton ($\tau^{(3)} = -1$):

$$\pm \frac{c^2 |g|^2 k^2}{2 V \mu^2 \omega_k^2},$$

which corresponds to a total scattering cross-section:

$$Q = \frac{|g|^4 k^4}{4 \pi \mu^4 (h \omega_k)^2} \tag{14.9}$$

This result differs by a factor $(k/\mu)^4$ from the corresponding result for scalar mesons (9.18). In the scalar case the cross-section decreases with increasing meson energy whereas the cross-section (14.9) increases without limit in the approximation discussed here (cf. §15).

The spin[1] of the nucleons does not enter into the coupling term (14.2). Therefore the spin remains unaffected in the transitions discussed so far. In particular, the nucleon spin does not appear in the nuclear forces (14.8). If we consider, for example, the collision of slow protons and neutrons the static potential will result in changes of momentum with exchange of charge (cf. §9) but

[1] We are not dealing here with the "isotopic" spin, i.e., the charge number, but with the angular momentum of the nuclear particles.

without exchange of spin. Exchange forces of that type are called "Heisenberg forces," since they were first used by Heisenberg[1] for the wave mechanical discussion of binding energies of compound nuclei. This is similar in the case of the scalar mesons which as particles without spin do not transmit any angular momentum. The nuclear forces are, however, predominately "Majorana forces," as we shall see in detail in §15, which in a collision can exchange the charge as well as the spin of the nucleons.[2] It is, therefore, of interest to discuss briefly besides (14.2) still another coupling term in which the nuclear spin enters.

It is known that the nucleon has the spin $\frac{1}{2}$. From its wave functions one can form a skew-symmetrical tensor (6-vector) $\zeta_{\mu\nu} = -\zeta_{\nu\mu}$. The ν, 4-components of this tensor vanish in the rest system, whereas its i, j-components ($i, j = 1, 2, 3$) represent the spin. With the help of this tensor and the "meson field strengths" $f_{\mu\nu}$ (12.3) we can construct an invariant interaction term:

$$L' = \frac{1}{2}\sum_{\mu\nu}\{\zeta_{\mu\nu} f_{\mu\nu} + \zeta^*_{\mu\nu} f^*_{\mu\nu}\}.$$

Assuming again that the nuclear particles are at rest, the terms $\mu = 4$ and $\nu = 4$ of the sum vanish and we have:

$$L' = \frac{1}{2}\sum_{ij}\{\zeta_{ij} f_{ij} + \zeta^*_{ij} f^*_{ij}\} = (\zeta \cdot \nabla\times \psi) + (\zeta^* \cdot \nabla\times \psi^*)$$

(ζ = vector with the components ζ_{23}, ζ_{31}, ζ_{12}). Since L' does not contain any time derivatives of the ψ_ν, and since ψ_4 does not enter, one obtains in the Hamiltonian the additional term $H' = -L'$:

$$H' = -\{\zeta\cdot(\nabla\times\psi) + \zeta^*\cdot(\nabla\times\psi^*)\}. \tag{14.10}$$

In the quantized theory, the components of ζ and ζ^* are matrices with respect to the spin indices of the nucleons and with respect to the charge numbers λ_n. We are setting in (14.10):

$$\zeta = \sum_n \sigma_n g_n \delta(x - x_n), \qquad \zeta^* = \sum_n \sigma_n g^*_n \delta(x - x_n), \tag{14.11}$$

where σ_n represents the spin vector of the nth nucleon, i.e., the vector σ_n has

[1] *Z. Phys.* 77, 1, 1932.

[2] Majorana, *Z. Phys.* 82, 137, 1933. For more exact definition of the Majorana force, cf. §15.

as components the Pauli spin-matrices $\sigma_n^{(j)}$ with the properties:

$$(14.12)\quad \sigma_n^{(j)} = \sigma_n^{(j)*};\ (\sigma_n^{(j)})^2 = 1;\ \sigma_n^{(2)}\sigma_n^{(3)} = -\sigma_n^{(3)}\sigma_n^{(2)} = i\,\sigma_n^{(1)}, \ldots, \ldots.$$

If we choose again in (14.11) for g_n and $\overset{*}{g}_n$ the isotopic spin-matrices (9.11,.15), then the equations (9.6, .7) hold again, i.e., the conservation of charge is guaranteed. With (5.5), (12.30, 12.31) the perturbation function can be written:

$$(14.13)\quad H' = -i\,V^{-1/2}\sum_k |k| \{(q_k^{(2)}\,\mathrm{e}_k^{(3)} - q_k^{(3)}\,\mathrm{e}_k^{(2)})\sum_n \sigma_n\,g_n\,e^{ikx_n} - (q_k^{(2)*}\,\mathrm{e}_k^{(3)} - q_k^{(3)*}\,\mathrm{e}_k^{(2)})\sum_n \sigma_n\,\overset{*}{g}_n\,e^{-ikx_n}\}.$$

This interaction affects thus only the transverse mesons ($r = 2, 3$).

In place of (14.5) we choose now:

$$(14.14)\quad S' = -\frac{c^2}{h}V^{-1/2}\sum_k \frac{|k|}{\omega_k^2}\{(p_k^{(2)}\,\mathrm{e}_k^{(3)} - p_k^{(3)}\,\mathrm{e}_k^{(2)})\sum_n \sigma_n\,\overset{*}{g}_n\,e^{-ikx_n} - (p_k^{(2)*}\,\mathrm{e}_k^{(3)} - p_k^{(3)*}\,\mathrm{e}_k^{(2)})\sum_n \sigma_n\,g_n\,e^{ikx_n}\},$$

so that again with (12.33):

$$[H^0, S'] = [H^{\mathrm{tr}}, S'] = -H'$$

In $\frac{1}{2}[H', S']$ the terms $n \neq n'$ yield:

$$H''_{(\neq)} = -c^2\sum_{n\neq n'} \overset{*}{g}_n\,g_{n'}\cdot\frac{1}{V}\sum_k \frac{k^2}{\omega_k^2}\{(\mathrm{e}_k^{(2)}\cdot\sigma_n)(\mathrm{e}_k^{(2)}\cdot\sigma_{n'}) + (\mathrm{e}_k^{(3)}\cdot\sigma_n)(\mathrm{e}_k^{(3)}\cdot\sigma_{n'})\}\cdot e^{ik(x_{n'}-x_n)}$$

$$= -c^2\sum_{n\neq n'} \overset{*}{g}_n\,g_{n'}\cdot\frac{1}{V}\sum_k \frac{1}{\omega_k^2}\{k^2(\sigma_n\cdot\sigma_{n'}) - (\sigma_n\cdot k)(\sigma_{n'}\cdot k)\}\cdot e^{ik(x_{n'}-x_n)},$$

or with (7.13):

$$(14.15)\quad H''_{(\neq)} = \sum_{n\neq n'} \overset{*}{g}_n\,g_{n'}\{(\sigma_n\cdot\sigma_{n'})\nabla_n^2 - (\sigma_n\cdot\nabla_n)(\sigma_{n'}\cdot\nabla_n)\}\,U(x_{n'} - x_n).$$

We shall not discuss here the terms $n = n'$. They describe among other effects a scattering of transverse mesons. The cross-section of this process shows the same energy-dependence as in the case of the longitudinal mesons [cf. (14.9)].[1]

[1] It may be mentioned that in the case of neutral vector mesons, where the g_n are no matrices but simple numerical coefficients, the scattering of the longitudinal mesons at the stationary proton-neutron disappears, but not that of the transverse mesons, since the three components of σ_n do not commute with each other.

For the case when only two nuclear particles are present, the formula (14.15) with $x_2 - x_1 = x$ can be written:

$$H''_{(\mp)} = \left(g_1 g_2^* + g_1^* g_2\right) \left\{(\sigma_1 \cdot \sigma_2)\nabla^2 - (\sigma_1 \cdot \nabla)(\sigma_2 \cdot \nabla)\right\} U(x).$$

The matrix $(g_1 g_2^* + g_1^* g_2)$ was already discussed in §9. Its eigen values are 0 (for $\lambda_1 = \lambda_2$) and $\pm|g|^2$ (for $\lambda_1 \neq \lambda_2$). To discuss the remaining factors we separate them as follows:

$$(14.16) \quad \begin{cases} H''_{(\mp)} = (g_1 g_2^* + g_1^* g_2)(A + B), \\ A = \frac{2}{3}(\sigma_1 \cdot \sigma_2)\nabla^2 U(x), \quad B = \left\{\frac{1}{3}(\sigma_1 \cdot \sigma_2)\nabla^2 - (\sigma_1 \cdot \nabla)(\sigma_2 \cdot \nabla)\right\} U(x). \end{cases}$$

The separation is chosen in such a way that B vanishes when averaged over all directions of x. One can also write for A according to (7.16):

$$A = \frac{2}{3}(\sigma_1 \cdot \sigma_2)\left[\mu^2 U(x) - \delta(x)\right];$$

here the term $\sim \delta(x)$ represents a point interaction between protons and neutrons, which is not compatible with the finite binding energy of the proton and neutron in the deuteron, and which must therefore be eliminated. This can be achieved by adding a suitable term in the Lagrangian $\left(L'' = \text{const.} \sum_{\mu\nu} \zeta^*_{\mu\nu} \zeta_{\mu\nu}\right)$. With this correction we have:

$$(14.17) \quad A = \frac{2}{3}\mu^2 (\sigma_1 \cdot \sigma_2)\, U(x).$$

The spin matrix $\sigma_1 \cdot \sigma_2 = \sum_j \sigma_1^{(j)} \sigma_2^{(j)}$ can easily be transformed into diagonal form, since the Pauli matrices $\sigma_n^{(j)}$ have the same representation as the isotopic spin matrices $\tau_n^{(j)}$. We have solved this problem already in §10 in the discussion of the matrix $(\tau_1 \cdot \tau_2)$ in the "symmetrical" theory, and we can simply take over the result by writing the spin coordinates instead of charge numbers λ_1, λ_2. $\sigma_1 \cdot \sigma_2$ has the three-fold eigen value $+1$ and the single eigen value -3; the respective eigen functions are symmetrical in the spin coordinates of the two particles for the case of the three eigen values $+1$, and antisymmetrical for the case of the eigen value -3. On the other hand the term B in (14.16) can be written as follows:

$$(14.18) \quad B = \left\{\frac{1}{3}(\sigma_1 \cdot \sigma_2) - \frac{(\sigma_1 \cdot x)(\sigma_2 \cdot x)}{r^2}\right\} r \frac{\partial}{\partial r}\left(\frac{1}{r}\frac{\partial U}{\partial r}\right).$$

If one introduces the resulting spin $s = \frac{1}{2}(\sigma_1 + \sigma_2)$ and its components in the x-direction $s_x = (s \cdot x)/r$, it follows on account of the relations (14.12):

$$s^2 = \frac{1}{2}\left\{3 + (\sigma_1 \cdot \sigma_2)\right\}, \qquad s_x^2 = \frac{1}{2}\left\{1 + \frac{(\sigma_1 \cdot x)(\sigma_2 \cdot x)}{r^2}\right\},$$

and hence:

$$(14.19) \quad B = 2\left\{\frac{1}{3}s^2 - s_x^2\right\} r \frac{\partial}{\partial r}\left(\frac{1}{r}\frac{\partial U}{\partial r}\right).$$

The eigen values of s^2 and s_x (for given $x = x_2 - x_1$) are:

$$s^2 = 0, \; s_x = 0 \quad \text{und} \quad s^2 = 2, \; s_x = 0, \; \pm 1,$$

which also determines the eigen values of B. In the same notations, A (14.17) can be written:

$$A = \mu^2 \cdot 2 \left\{ \frac{2}{3} s^2 - 1 \right\} U. \tag{14.20}$$

The two interactions (14.2) and (14.10) are the only ones compatible with the relativistic postulates and which are linear in the meson field variables. They can, of course, also be combined linearly [H' = sum of (14.2) and (14.10)], and the coupling parameters $|g|$ in both terms can be chosen independently. In this case longitudinal as well as transverse mesons are coupled with the nuclear particles: both are scattered and both transmit nuclear forces of the type (14.8) or (14.15 to 14.20).

As we have done in §10 for the scalar field, we can finally construct a "symmetrical theory" by using in addition a neutral vector meson field.[1] We shall not discuss the corresponding interaction terms here. As to the nuclear forces, it follows again that in (14.8, 14.15) the isotopic spin operators:

$$\overset{*}{g}_n g_{n'} + \overset{*}{g}_{n'} g_n = \frac{1}{2} |g|^2 \left\{ \tau_n^{(1)} \tau_{n'}^{(1)} + \tau_n^{(2)} \tau_{n'}^{(2)} \right\}$$

[cf. (9.22)] will be replaced by:

$$\frac{1}{2} |g|^2 (\tau_n \cdot \tau_{n'}) = \frac{1}{2} |g|^2 \left\{ \tau_n^{(1)} \tau_{n'}^{(1)} + \tau_n^{(2)} \tau_{n'}^{(2)} + \tau_n^{(3)} \tau_{n'}^{(3)} \right\}$$

The most general static interaction potential between two proton-neutrons can thus, according to (14.8, 14.16, 14.17), be written in the symmetrical vector theory:

$$V_{12} = (\tau_1 \cdot \tau_2) \left\{ \left[|g|^2 + \frac{2}{3} |f|^2 (\sigma_1 \cdot \sigma_2) \right] U(x) + \frac{1}{\mu^2} |f|^2 \cdot B \right\}, \tag{14.21}$$

where $|g|$ and $|f|$ are two independent coupling parameters. These nuclear forces are "charge-independent" in the sense discussed earlier (in §10): the force between a neutron and a proton, which are symmetrical in the charge numbers [$(\tau_1 \cdot \tau_2 = 1)$], is the same as that between two neutrons or two protons.

§ 15. Meson Theory and Nuclear Forces

We have discussed in §§9, 10, and 14 the Yukawa-Kemmer theory of interaction between mesons and nuclear particles primarily for the purpose of illustrating the typical methods of calculation which are used in the applications of the quantized theory. In addition, this theory deserves special interest as one of the most frequently discussed field theories of nuclear forces. For this reason we shall discuss here briefly to what extent this theory explains the known facts of nuclear physics.

[1] Kemmer, *loc. cit.* in §10.

We shall start out with the results of the phenomenological nuclear theory. In this theory one assumes the existence of nuclear forces with such properties that the stability of the known nuclei results from the wave mechanical calculations based on these forces. The resulting semi-empirical knowledge of the nuclear forces is in many respects still incomplete and not quite independent of arbitrary hypotheses. It rests, however, on safer ground than a field theory where detailed assumptions must be made about the nature of the field transmitting the nuclear forces and on its interaction with the nuclear particles.[1] It seems reasonable, therefore, to test the field theory by comparing it with the phenomenological theory.

In this theory one assumes with Heisenberg[2] that forces between two particles (protons or neutrons) inside the atomic nucleus can, at least in first approximation, be considered as static central forces, i.e., they can be represented by a potential function $J(r)$ (r = distance of particles). In order to obtain the saturation property of the nuclei, it is assumed that these forces shall be so-called exchange forces. Heisenberg assumed originally that such forces of non-electrical nature exist essentially only between protons and neutrons. Scattering experiments of protons on protons at energies of the order of magnitude of 1 m.e.v. showed, however, that an additional force besides the electrostatic Coulomb force must exist between two protons.[3] The quantitative wave mechanical discussion led to the conclusion that this nuclear proton-proton force in the ^1S state is equal or almost equal to the proton-neutron force, a fact which speaks for the hypothesis of the "charge independence" of nuclear forces.[4] On account of this fact the charge independence is used in almost all theories of nuclear forces. For charge-independent forces there exist four independent forms of the interaction potential V, corresponding to the different exchange character of the forces:

Wigner force: $$V = J(r),$$

Bartlett force: $$V = \frac{1 + (\sigma_1 \cdot \sigma_2)}{2} J(r),$$

Heisenberg force: $$V = -\frac{1 + (\tau_1 \cdot \tau_2)}{2} J(r),$$

Majorana force: $$V = -\frac{1 + (\sigma_1 \cdot \sigma_2)}{2} \frac{1 + (\tau_1 \cdot \tau_2)}{2} J(r).$$

According to the discussions in §14, the operators $\frac{1 + (\sigma_1 \cdot \sigma_2)}{2}$ and $\frac{1 + (\tau_1 \cdot \tau_2)}{2}$ have the eigen values ± 1[5] and the corresponding eigen functions are symmetrical or antisymmetrical in the spin or the charge coordinates. On account of this symmetry

[1] As a forerunner to the meson theory we mention the β-theory of the nuclear forces according to which the transmitting field is an electron-neutrino field. (Tamm and Jwanenko, *Nature* *133*, 981, 1934). However, the coupling of this field with the nuclear particles is too weak on account of the long lifetime of the β-emitter.

[2] *Z. Phys.* *77*, 1, 1932.

[3] Tuve, Heydenburg, and Hafstad, *Phys. Rev.*, *50*, 806, 1936.

[4] Breit, Condon and Present, *Phys. Rev.* *50*, 825, 1936; Breit and Feenberg, *ibid.* *50*, 850, 1936.

[5] The eigen values of $(\sigma_1 \cdot \sigma_2)$ and $(\tau_1 \cdot \tau_2)$ are $+1$ (threefold) and -3 (single); cf. §§10 and 14.

property one can also say: $\frac{1 + (\sigma_1 \cdot \sigma_2)}{2}$ or $\frac{1 + (\tau_1 \cdot \tau_2)}{2}$ is the operator which interchanges the spin or the charge coordinates respectively of the particles 1 and 2, since this operator is equivalent to multiplication with $+1$ or -1 if applied to the symmetrical or antisymmetrical function. The operator

$$\left(-\frac{1 + (\sigma_1 \cdot \sigma_2)}{2} \quad \frac{1 + (\tau_1 \cdot \tau_2)}{2}\right)$$

of the Majorana force is equivalent to the operator which exchanges the space coordinates alone. This is so because the wave function is, according to the exclusion principle of Pauli, antisymmetrical in all the coordinates (position, spin, and charge coördinates) of any two nucleons.

If one assumes that all four types of forces exist and that they exist with the same r-dependence, one arrives at the generally used expression for the potential:

$$V = \{c + c_\sigma (\sigma_1 \cdot \sigma_2) + c_\tau (\tau_1 \cdot \tau_2) + c_{\sigma\tau} (\sigma_1 \cdot \sigma_2)(\tau_1 \cdot \tau_2)\} J(r), \tag{15.1}$$

where c, c_σ, c_τ, $c_{\sigma\tau}$ are four independent constants. As to the choice of the function $J(r)$, it is essential that a "finite" range should be assigned to these forces: $J(r)$ shall be different from zero only for $r \lesssim a$. The most frequently used expressions for $J(r)$ are:

$$J(r) = e^{-\frac{r}{a}}, \qquad J(r) = e^{-\left(\frac{r}{a}\right)^2}, \qquad J(r) = \begin{cases} 1 \text{ for } r < a, \\ 0 \text{ for } r > a; \end{cases}$$

Recently the Yukawa potential:

$$J(r) = \frac{a}{r} e^{-\frac{r}{a}}$$

has also been considered. The main results are to a large extent independent of the special type of function used.

Two first conditions for the constants in (15.1) are obtained from the two-body problem. The ground state of the deuteron is a ^3S-state. The eigen function is thus symmetrical in the space and spin coordinates of the proton and neutron and, according to the exclusion principle, antisymmetrical in the charge coordinates: $\sigma_1 \cdot \sigma_2 = 1$, $\tau_1 \cdot \tau_2 = -3$. From the known binding energy of the deuterons one can deduce the numerical value of:

$$c + c_\sigma - 3\,c_\tau - 3\,c_{\sigma\tau} \; (< 0)$$

for any special type of function $J(r)$. The energy of the deuteron in the ^1S-state, is also well known. This state is responsible for the strong scattering of slow neutrons on hydrogen. It also plays an important part in the theory of the photoelectric disintegration of the deuteron, as well as in the inverse process, the proton-neutron recombination under γ-emission. For this state the eigen function is symmetrical in the space and charge coordinates, and antisymmetrical in the spin coordinates $\sigma_1 \cdot \sigma_2 = -3$, $\tau_1 \cdot \tau_2 = 1$. From its known energy one can determine the constant:

$$c - 3\,c_\sigma + c_\tau - 3\,c_{\sigma\tau}\ (< 0)$$

This same ^1S-interaction should be active for the proton-proton scattering, if the forces are really charge-independent. The fact that this is really so—in the case of the Yukawa potential even inside the limits of error[1]—is the main foundation for the hypothesis of the independency of charge.

While the three- and four-body problems (H^3, He^3, He^4) essentially allow only to deduce the range a of forces, new statements about the constants c can be derived from the theory of heavy nuclei. One uses here the Hartree-Fock method of approximation. In the first approximation two kinds of terms appear in the potential energy of the nucleus:

$$V_1 = \int dx_1 \int dx_2\, \varrho\,(x_1, x_1)\, \varrho\,(x_2, x_2)\, J\,(|x_2 - x_1|)\cdot\text{const.},$$
$$V_2 = \int dx_1 \int dx_2\, \varrho\,(x_1, x_2)\, \varrho\,(x_2, x_1)\, J\,(|x_2 - x_1|)\cdot\text{const.},$$

where $\rho(x_1, x_2)$ is the mixed density function (density matrix):

$$\varrho\,(x_1, x_2) = \sum_n u_n^*\,(x_1)\, u_n\,(x_2)$$

(u_n = space eigen function of the nuclear particles in the occupied states). The energy term V_1, which depends on the ordinary particle density $\rho(x, x)$, must be positive in all cases. Otherwise, very small and stable nuclei could exist—in contradiction to experience. It follows that V_2 must be negative, larger in magnitude than V_1. Since these forces have saturation properties, nuclear volume and binding energy are about proportional to the number of particles or to the atomic weight, as is demanded by experiment. The condition, that V_1 should be positive for all conceivable nuclei, yields several inequalities for the coefficients c in (15.1), which are called "saturation conditions." They are especially important because their validity does not depend on the Hartree approximation. The reason is that this approximation can yield for the ground state only energy values which are too high. Applied to such nuclei, in which all states of the particles are always realized with both spin orientations, one can form the mean over all spin quantum numbers, whereby the terms in (15.1) which are multiplied by $\sigma_1 \cdot \sigma_2$ are cancelled, since $\sigma_1 \cdot \sigma_2$ has the trace 0. One can form in the same way the mean over the charge quantum numbers if the same number of protons and neutrons are present and if all spatial states are occupied in the same way by protons and neutrons; in that case the terms proportional to $\tau_1 \cdot \tau_2$ in V_1 make no contribution and the saturation condition is:[2]

$$c \geqq 0.$$

To avoid on the other hand the existence of stable nuclei, with only protons or neutrons present, whose eigen functions therefore would have to be symmetrical in the charge numbers of all particles, one must demand[3] that V_1 be positive, if in

[1] Hoisington, Share, and Breit, *Phys. Rev. 56*, 884, 1939.

[2] Breit and Feenberg, *Phys. Rev. 50*, 850, 1936.

[3] Kemmer, *Nature 140*, 192, 1937.

(15.1) $\tau_1 \cdot \tau_2 = 1$ and $\sigma_1 \cdot \sigma_2$ (mean value) $= 0$:

$$c + c_\tau \geqq 0.$$

Similarly, on account of the symmetry of the equations with respect to spin and charge numbers:

$$c + c_\sigma \geqq 0$$

This is the condition that no very stable nuclei should exist with all spins parallel.

If one combines these inequalities with the more qualitative statements, which can be obtained from the Hartree approximation by comparison with the known properties of nuclei, one finds best agreement with the facts if the constants c and c_σ are put equal to zero. In this case the data on deuterons yield values for the two other constants c_τ and $c_{\sigma\tau}$; they are positive and have approximately the ratio $c_\tau : c_{\sigma\tau} \cong 1 : 2$.

With $c = c_\sigma = 0$, $c_\tau > 0$, $c_{\sigma\tau} > 0$, the interaction (15.1) is the same as that of formula (14.21), if one neglects the B-term in the latter. As far as this is justified, the symmetrical vector meson theory seems to be satisfactory. If one adds to (15.1) a term const. $(\tau_1 \cdot \tau_2)B$, then this term is of no great importance in the theory of heavy nuclei, since it is equal to zero in the mean over all spin orientations as well as in the mean over all orientations of the vector $x = x_2 - x_1$. The influence of this term on the deuteron must be significant, however. According to (14.19) B depends on the orientation of the radius vector x with respect to the deuteron spin s. From this fact it follows that we have a spin-orbit coupling and the eigen function loses its spherical symmetry in the deuteron ground state. The ^3S-function gets a "^3D-admixture." Consequently the spatial distribution of the electric charge in the deuteron also shows a deviation from the spherical symmetry of such a kind that an electrical quadrupole moment results. This prediction of the field theory has been verified experimentally: Kellogg, Rabi, Ramsey, and Zacharias[1] have observed transitions in heavy molecules of hydrogen (HD and D_2) which were induced by radio frequency fields, and they have shown that the deuteron has in fact an electric quadrupole moment. Quantitatively, however, difficulties arise. First there are difficulties of a principle nature. If one substitutes in B (14.19) for U the Yukawa potential, then B increases for $r \to 0$ like r^{-3}, and such a strong singularity is not compatible with the finite binding energy of the deuteron. Hence one is forced to cut off, or at least to weaken, the singularity by introducing a form factor. Calculating with a potential function U, changed in this way, one obtains from relation (14.19) the ^3D-admixture to the ^3S-Schrödinger function as small as demanded (on account of the smallness of the amplitude of the D-wave function in the domain of the S-function). The calculated quadrupole moment has thus the correct order of magnitude. There is however a discrepancy with regard to the sign of the quadrupole moment. The theoretical charge distribution would be an ellipsoid of revolution, flattened in the direction of the spin axis[2] whereas the experimental results indicate an ellipsoid with its major axis in the spin direction.

This failure of the vector theory to give the correct sign of the quadrupole moment

[1] *Phys. Rev.* 57, 677, 1940.

[2] This is based on the fact that the B-term in (14.21) in case of the deuteron ground state $[\tau_1 \cdot \tau_2 = -3]$ represents a repelling potential according to (14.19), if the space vector x is parallel to the spin s, but it gives an attractive potential, if x is normal to s. Cf. Bethe, *Phys. Rev.* 57, 390, 1940.

of the deuteron can hardly be due to the cutting off of the potential function at small distances. We may therefore conclude that the vector theory cannot account for the known properties of nuclear forces.

This difficulty can be remedied if one introduces besides the vector field a pseudoscalar meson field.[1] In vacuum the pseudoscalar field satisfies exactly the same equations as the scalar field discussed in §8. The interaction with the heavy particle is, however, different on account of the different transformation property of a pseudoscalar under reflections. The calculation for the nuclear force in this case gives two additional terms in the interaction operator, one of which is exactly of the same form as the last term in (14.21) but with opposite sign. If the mass of the vector meson is chosen larger than the mass of the pseudoscalar meson, then these two terms do not cancel but combine in such a way that a tensor force of the correct sign results. Moreover the r^{-3} singularity at the origin cancels and the potential of this combined term behaves for small distances like r^{-1}. This seems very satisfactory, but a more detailed quantitative investigation of this mixed theory shows that the tensor force which results from this term can account for only about half the observed value of the quadrupole moment of the deuteron.[2,3]

In judging the meson theory of nuclear forces another difficulty must be discussed. It should be emphasized that the formula (14.21) represents only a first approximation in a power series in rising powers of the coupling constants f and g. The calculation of higher order terms with the ordinary methods of perturbation theory leads to divergent results. In order to estimate the magnitudes of these terms it is therefore necessary either to use an extended source function for the nucleons or to introduce some kind of subtraction formalism to eliminate the divergent terms. In the first case the relativistic invariance of the theory is of course abandoned and moreover it seems impossible to reconcile the conditions for "weak coupling" with any reasonable assumption about the cut-off radius. In the second case it is possible to retain the relativistic invariance and the "weak coupling" condition.[4] But even under these favorable conditions the theory is not satisfactory, for instance, the non-static forces having inadmissable singularities at the origin.[5]

Besides nuclear forces, the meson theory has been claimed to explain the anomalous magnetic moments of the proton and the neutron. If nucleons interact with the electromagnetic field, they would have a magnetic moment provided they satisfy the Dirac wave equation. This would be one nuclear magneton ($eh/2Mc$) for the proton and 0 for the neutron. Through the coupling with the charged meson field an additional dipole interaction with an external magnetic field can occur, that is, the meson field surrounding the nucleon can make a contribution to its magnetic moment. In order to calculate this effect one must combine the coupling with the nucleons and the electromagnetic field used in §§13 and 14 and carry out a third order perturbation calculation.[6] Since the resulting integrals

[1] C. Møller and L. Rosenfeld, *Kgl. Danske Vid. Selskab, Math.-fys. Medd.* *17*, No. 8, 1940. J. Schwinger, *Phys. Rev.* *61*, 387, 1942.

[2] J. M. Jauch and Ning Hu, *Phys. Rev.* *65*, 289, 1944.

[3] For a complete up to date discussion of the present situation in the meson theory of nuclear forces, cf. G. Wentzel, *Rev. Modern Phys.* *19*, 1, 1947.

[4] W. Pauli, *Phys. Rev.* *64*, 332, 1942.

[5] N. Hu, *ibid.* *67*, 339, 1945.

[6] Fröhlich, Heitler, and Kemmer, *Proc. Roy. Soc. London* *166*, 154, 1938.

in momentum space diverge, the theory does not tell anything about the numerical values of the magnetic moments; however, it explains qualitatively the experimental fact that the proton moment differs from 1 and the neutron moment from 0,[1] and in addition it yields a statement concerning the relationship of the two moments. Since the meson fields in the neighborhood of a proton and a neutron differ only in the signs of their electrical charge density, we obtain in the lowest (third) approximation of the perturbation theory the excess of the proton moment over a nuclear magneton, equal in its amount to the neutron moment, whereas the signs are opposite (i.e., for parallel mechanical spin moments the respective magnetic moments are antiparallel). The predictions of the theory, which of course is only approximative in character, have been confirmed by measurements.[2] In the deuteron ground state where the spins of proton and neutron have the same direction, both contributions of the meson field cancel each other approximately, so that the deuteron moment is equal to about 1 nuclear magneton.[3]

Supposing that the physical ideas underlying the meson theories are basically correct—notwithstanding their mathematical shortcomings—we are confronted with the crucial problem to identify the particle which we hypothetically introduced and called "meson," with a real particle, observed in nature. The mass of the particle in question can be estimated from the range of the nuclear forces, according to the relation $1/\mu = h/mc$, discussed in §7. The most exact experimental value for $1/\mu$ is obtained from the wave mechanical discussion of the proton-proton scattering. If one chooses for $J(r)$ in (15.1) the Yukawa potential U, the energy-dependence and the directional distribution of the scattering are obtained most accurately for $1/\mu = 1.2 \times 10^{-13}$ cm.[4] The corresponding particle should then have a mass of roughly 300 electron masses. This suggested the identification with the cosmic ray "meson." Recent experiments have shown, however, that the particle known from cosmic radiation (at sea level), besides having rather too low a mass (about 200 electron masses only), does not interact strongly enough with atomic nuclei and therefore cannot be the agent responsible for the nuclear forces. A heavier meson has been discovered in the cosmic radiation by Occhialini, Powell, Lattes, and Muirhead,[5] and the same particle was also artificially produced in the Berkeley cyclotron.[6] Its mass value as well as its strong interaction with nuclei (production of "stars") suggests that it might be identified with the meson of the Yukawa theory. This seems to be a promising outlook, although much further experimental work will be needed to decide this important question.

---

[1] Frisch and Stern, *Z. Phys.* *85*, 4, 1933; Kellogg, Rabi, Ramsey, and Zacharias, *Phys. Rev.* *56*, 728, 1939; Millman and Kusch, *Phys. Rev.* *60*, 91, 1941; Alvarez and Bloch, *Phys. Rev.* *57*, 111, 1940. According to Rabi and his co-workers the magnetic moment of the proton is 2.7896 ± 0.0008 nuclear magnetons. The best value of the neutron moment is due to W. R. Arnold and A. Roberts, *Phys. Rev.* *71*, 878, 1947. They find $\mu = -1.9103 \pm 0.0012$.

[2] Cf. last footnote.

[3] $+0.8567 \pm 0.003$ nuclear magnetons according to the measurements by Arnold and Roberts. According to Rarita and Schwinger, *Phys. Rev.* *59*, 436, 1941, about 3.9% admixture of ^3D function is to be expected from the known value of the quadrupole moment in the deuteron ground state. The sum of proton and neutron moment is 0.8793 and the excess moment over the deuteron 0.0228 is outside the experimental error and is in satisfactory agreement with the additional moment due to the ^3D part of the wave function.

[4] Hoisington, Share, and Breit, *Phys. Rev.* *56*, 884, 1939.

[5] *Nature* *159*, 186, 694 (1947); *160*, 453, 486 (1947).

[6] E. Gardner and C. M. G. Lattes, *Science* *107*, 270 (1948).

Chapter IV

Quantum Electrodynamics

§ 16. The Electromagnetic Field in Vacuum

The electromagnetic field differs from the vector meson field in two respects: first, it is a real, electrically neutral field; second, the parameter μ vanishes and with it the rest mass $m = \mu h/c$ of the corresponding corpuscles, light quanta, or photons. To underline the analogy to the meson field, as far as it exists, we denote components of the electromagnetic field strengths by $f_{\mu\nu}$:

$$f_{4j} = -f_{j4} = i\,\mathfrak{E}_j, \quad f_{23} = -f_{32} = \mathfrak{H}_1, \; \ldots, \; \ldots. \tag{16.1}$$

According to Maxwell's equations in vacuum the skew-symmetrical tensor $f_{\mu\nu}$ can be represented as curl of a 4-potential ψ_ν:

$$f_{\mu\nu} = \frac{\partial \psi_\nu}{\partial x_\mu} - \frac{\partial \psi_\mu}{\partial x_\nu}, \tag{16.2}$$

and its divergence vanishes:

$$\sum_\mu \frac{\partial f_{\mu\nu}}{\partial x_\mu} = 0. \tag{16.3}$$

This agrees with the equations (12.3, 12.4), if we set in them $\mu = 0$.

The equations (16.2) for the field $f_{\mu\nu}$ do not determine ψ_ν. In fact the field variables $f_{\mu\nu}$ are invariant under the gauge transformation:[1]

$$\psi_\nu \rightarrow \psi_\nu + \frac{\partial \Lambda}{\partial x_\nu}, \tag{16.4}$$

where Λ is an arbitrary scalar function. Since all electromagnetic effects are determined by the values of the field variables $f_{\mu\nu}$ alone, one must formulate the theory in such a way that all measurable quantities are defined as gauge-

[1] Cf. footnote 1, p. 68.

invariant. This is a further difference between the electromagnetic field and the vector meson field. It is connected, of course, with the fact that $\mu = 0$, for in case $\mu \neq 0$ there exists no group of gauge transformations.

If one chooses the Lagrangian, analogous to (12.5):

$$L = -\frac{1}{4}\sum_{\mu\nu}\left(\frac{\partial\psi_\nu}{\partial x_\mu} - \frac{\partial\psi_\mu}{\partial x_\nu}\right)^2,$$

the field equations (12.6) follow with $\mu = 0$:

$$\sum_\mu \frac{\partial}{\partial x_\mu}\left(\frac{\partial\psi_\nu}{\partial x_\mu} - \frac{\partial\psi_\mu}{\partial x_\nu}\right) = 0,$$

in accordance with (16.2, 16.3). Corresponding to (12.8) $\pi_\nu^* = f_{\nu 4}/ic$, and hence again $\pi_4 = 0$. Here it is not possible to eliminate ψ_4, as in §12, since the method there depends on the condition $\mu \neq 0$. We shall choose in the following a canonical formalism which avoids the difficulty of the identical vanishing of π_4.[1] Another method has been proposed by Heisenberg and Pauli.[2]

Let the Lagrangian be:

$$L = -\frac{1}{4}\sum_{\mu\nu}\left(\frac{\partial\psi_\nu}{\partial x_\mu} - \frac{\partial\psi_\mu}{\partial x_\nu}\right)^2 - \frac{1}{2}\left(\sum_\mu \frac{\partial\psi_\mu}{\partial x_\mu}\right)^2. \tag{16.5}$$

Then the field equations (1.2) are:

$$\frac{\partial L}{\partial\psi_\nu} - \sum_\mu \frac{\partial}{\partial x_\mu}\frac{\partial L}{\partial\dfrac{\partial\psi_\nu}{\partial x_\mu}} = \sum_\mu \frac{\partial}{\partial x_\mu}\left(\frac{\partial\psi_\nu}{\partial x_\mu} - \frac{\partial\psi_\mu}{\partial x_\nu}\right) + \frac{\partial}{\partial x_\nu}\sum_\mu \frac{\partial\psi_\mu}{\partial x_\mu} \tag{16.6}$$

$$= \sum_\mu \frac{\partial^2}{\partial x_\mu^2}\psi_\nu \equiv \Box\,\psi_\nu = 0.$$

For the field variables $f_{\mu\nu}$ defined by (16.2) these equations imply:

$$\sum_\mu \frac{\partial f_{\mu\nu}}{\partial x_\mu} + \frac{\partial\chi}{\partial x_\nu} = 0, \tag{16.7}$$

where:

$$\chi = \sum_\mu \frac{\partial\psi_\mu}{\partial x_\mu}. \tag{16.8}$$

---

[1] Fermi, *Atti accad. Lincei* *9*, 881, 1929, and *12*, 431, 1930; *Rev. Modern Phys.* *4*, 87, 1932, part III. We follow here a paper by Dirac, Fock, and Podolsky: *Physik. Z. Sowjetunion* *2*, 468, 1932.

[2] *Z. Phys.* *56*, 1, 1929, and *59*, 168, 1930.

In order for (16.7) to agree with Maxwell's equations, χ must be constant in space and time. It is sufficient, however, as we shall see, to require that χ and $\partial\chi/\partial t$ should vanish everywhere for $t = 0$:

$$\chi = 0 \quad \text{and} \quad \frac{\partial \chi}{\partial t} = 0 \quad \text{for } t = 0. \tag{16.9}$$

In order to see this we assume that χ is expanded in powers of t for an arbitrary position:

$$\chi = \chi_{t=0} + \frac{1}{1!}\, t \left(\frac{\partial \chi}{\partial t}\right)_{t=0} + \frac{1}{2!}\, t^2 \left(\frac{\partial^2 \chi}{\partial t^2}\right)_{t=0} + \cdots;$$

the first two terms of this series vanish according to (16.9). Since according to (16.6, 16.8) $\Box\, \chi = 0$, $\partial^2\chi/\partial t^2 = c^2\nabla\chi$ and $\partial^3\chi/\partial t^3 = c^2\nabla^2\partial\chi/\partial t$ vanish for $t = 0$, and also $\partial^4\chi/\partial t^4 = c^4\nabla^4\chi$ and all higher terms. Thus χ vanishes identically in the domain of convergence of the series, which means that χ vanishes identically everywhere:

$$\chi = 0. \tag{16.10}$$

The addition of the initial conditions (16.6) thus has the effect that the Maxwell fields are singled out from the totality of solutions of the more general fields satisfying (16.6).[1]

To transfer this line of thought into the quantum theory we must first change to the Hamiltonian formalism. According to (16.5) the fields, which are canonical conjugates to the potentials ψ_ν, are:

$$\pi_\nu = \frac{1}{ic} \frac{\partial L}{\partial \dfrac{\partial \psi_\nu}{\partial x_4}} = \frac{1}{ic}\left\{\frac{\partial \psi_4}{\partial x_\nu} - \frac{\partial \psi_\nu}{\partial x_4} - \delta_{\nu 4} \sum_\mu \frac{\partial \psi_\mu}{\partial x_\mu}\right\},$$

hence according to (16.1, 16.2, and 16.8):

$$\pi_j = \frac{1}{ic} f_{j4} = -\frac{1}{c}\mathfrak{E}_j, \quad \pi_4 = \frac{i}{c}\sum_\mu \frac{\partial \psi_\mu}{\partial x_\mu} = \frac{i}{c}\chi. \tag{16.11}$$

[1] The group of gauge transformations of the potential is restricted by the subsidiary condition (16.10): the scalar function Λ in (16.4) must satisfy the wave equation $\Box\Lambda = 0$. Any more general potential field can easily be made to satisfy the condition (16.10) by subjecting it to a suitable gauge transformation.

Solving these with respect to $\dot{\psi}_\nu$ we obtain:

$$\dot{\psi}_j = c^2 \pi_j + i c \frac{\partial \psi_4}{\partial x_j}, \qquad \dot{\psi}_4 = c^2 \pi_4 - i c \sum_j \frac{\partial \psi_j}{\partial x_j},$$

or in vector notation:

$$\dot{\psi} = c^2 \pi + i c \nabla \psi_4, \qquad \dot{\psi}_4 = c^2 \pi_4 - i c \nabla \cdot \psi. \tag{16.12}$$

We obtain now for the Hamiltonian (1.5):

$$H = \frac{1}{2} c^2 (|\pi|^2 + \pi_4^2) + \frac{1}{2} |\nabla \times \psi|^2 + i c \{ \pi \cdot \nabla \psi_4 - \pi_4 \nabla \cdot \psi \}. \tag{16.13}$$

The canonical commutation rules (1.7) can now be taken over without change:

$$\begin{cases} [\psi_\nu (x), \psi_{\nu'} (x')] = [\pi_\nu (x), \pi_{\nu'} (x')] = 0, \\ [\pi_\nu (x), \psi_{\nu'} (x')] = \frac{h}{i} \delta_{\nu\nu'} \delta(x - x'). \end{cases} \tag{16.14}$$

The canonical field equations which follow for $\dot{\psi}_\nu$, agree with (16.12).[1] On the other hand it follows for $\dot{\pi}_\nu$:

$$\dot{\pi} = - \nabla \times (\nabla \times \psi) - i c \nabla \pi_4, \qquad \dot{\pi}_4 = i c \nabla \cdot \pi. \tag{16.15}$$

Eliminating π and π_4 from these equations, one finds again the field equations (16.6):

$$\ddot{\psi} = c^2 \nabla^2 \psi, \qquad \ddot{\psi}_4 = c^2 \nabla^2 \psi_4. \tag{16.16}$$

In the same way the elimination of ψ and ψ_4 yields:

$$\ddot{\pi} = c^2 \nabla^2 \pi, \qquad \ddot{\pi}_4 = c^2 \nabla^2 \pi_4. \tag{16.17}$$

On account of the equations (4.17) (with $\mu = 0$) which are satisfied according to (16.16), the time-dependent operators:

$$\psi_\nu (x, t) = e^{\frac{it}{h} H} \psi_\nu (x) e^{-\frac{it}{h} H} \tag{16.18}$$

satisfy the commutation rules (4.28), where the $d^{(1)}_{\sigma\sigma'}$ [according to (4.21) and (16.12)] are equal to $c^2 \delta_{\sigma\sigma'}$:

$$[\psi_\nu (x, t), \psi_{\nu'} (x', t')] = \frac{h}{i} c^2 \delta_{\nu\nu'} \cdot D (x - x', t - t'). \tag{16.19}$$

[1] Cf. the analogous calculations in §12, which lead to the equations (12.17, 12.18).

These commutation rules are according to §4 equivalent to the canonical relations (16.14), and they are evidently invariant under Lorentz transformations. D represents here again the invariant function (4.25) but with the special parameter value $\mu = 0$:

$$D(x,t) = \frac{1}{(2\pi)^3}\int dk\, e^{ikx}\frac{\sin(tc\,|k|)}{c\,|k|} \tag{16.20}$$

(invariant D-function of Jordan and Pauli).[1] If we introduce further the time-dependent field operators by:

$$f_{\mu\nu}(x,t) = e^{\frac{it}{h}H} f_{\mu\nu}(x)\, e^{-\frac{it}{h}H}$$
$$= \frac{\partial}{\partial x_\mu}\psi_\nu(x,t) - \frac{\partial}{\partial x_\nu}\psi_\mu(x,t),$$

we obtain for them from (16.12) the commutation rules:[2]

$$[f_{\mu\nu}(x,t), f_{\mu'\nu'}(x',t')] = \frac{h}{i}c^2\left\{\delta_{\nu\nu'}\frac{\partial^2}{\partial x_\mu\,\partial x_{\mu'}} - \delta_{\nu\mu'}\frac{\partial^2}{\partial x_\mu\,\partial x_{\nu'}}\right.$$
$$\left. - \delta_{\mu\nu'}\frac{\partial^2}{\partial x_\nu\,\partial x_{\mu'}} + \delta_{\mu\mu'}\frac{\partial^2}{\partial x_\nu\,\partial x_{\nu'}}\right\} D(x-x', t-t').$$

The field which is defined by the Hamiltonian (16.13), is still too general, like the corresponding classical field. In fact the canonical field equations (16.12, 16.15) are equivalent to the equation (16.7). In order to single out the Maxwell fields, we shall impose initial conditions such that $\chi = 0$ [or according to (16.11) $\pi_4 = 0$] in the whole space-time continuum. In quantum mechanics it is, however, impossible to put $\pi_4 \equiv 0$ because this would violate the commutation rules (16.14). It is, however, sufficient to demand that the operator π_4 or χ, applied to the Schrödinger function, should give zero.

Let us consider for more detailed explanation the Schrödinger function of the system: $F = F(t, q)$. We write here q for the totality of the variables of the system [$q_i = \psi_\sigma^{(s)}$ in the notation of §1; after Fourier analysis according to (5.1) we can instead use the Fourier coefficients $q_{\sigma,k}$ or $\overset{(r)}{q}_k$, see below]. F is a solution of the Schrödinger equation:

$$\left(\frac{h}{i}\frac{\partial}{\partial t} + H\right)F(t,q) = 0,$$

[1] Cf. footnote 1, p. 26.

[2] The corresponding uncertainty relations (4.34) were discussed by Bohr and Rosenfeld (cf. footnote on p. 26) with regard to the hypothetical experiments to measure the field strengths in two space-time domains with greatest possible accuracy.

where H is the integral Hamiltonian operator given by (16.13) $\left(H = \int dx\, \mathbf{H}, \partial H/\partial t = 0\right)$. If F at the time $t = 0$ is given as function of the q, one obtains its value at the time t by applying the operator $e^{-\frac{it}{h}H}$ to $F(0, q)$, thus:

$$F(t, q) = e^{-\frac{it}{h}H} F(0, q); \tag{16.21}$$

because this function satisfies the Schrödinger equation and it reduces for $t = 0$ into $F(0, q)$. The above-mentioned postulate may now be written:

$$\chi(x)\ F(t, q) = 0 \quad \text{oder} \quad \pi_4(x)\ F(t, q) = 0. \tag{16.22}$$

If we operate with $e^{-\frac{it}{h}H}$ on the function to the left side of the above equation and if $F(t, q)$ is expressed according to (16.21) by $F(0, q)$ the equation (16.22) expresses that the time-dependent operator:

$$\chi(x, t) = e^{\frac{it}{h}H}\ \chi(x)\, e^{-\frac{it}{h}H}, \tag{16.23}$$

applied to $F(0, q)$, reduces this function to zero:

$$\chi(x, t) \cdot F(0, q) = 0. \tag{16.24}$$

We may now expand the operator $\chi(x, t)$, as before in the classical case, into a Taylor series:

$$\chi(x, t) = \sum_{n=0}^{\infty} \frac{1}{n!} t^n \overset{(n)}{\chi}(x) \tag{16.25}$$

$$= \chi(x) + \frac{1}{1!} t\, \dot{\chi}(x) + \frac{1}{2!} t^2\, \ddot{\chi}(x) + \dots$$

[cf. (4.7) and its footnote].

It remains to be shown that the postulate (16.22 or 16.24) is equivalent to an initial condition and hence can be satisfied. According to the model of the classical initial condition (16.9), we shall require:

$$\chi(x)\ F(0, q) = 0, \qquad \dot{\chi}(x)\ F(0, q) = 0, \tag{16.26}$$

which implies according to (16.11 and 16.15):

$$\pi_4(x)\ F(0, q) = 0, \qquad \operatorname{div} \pi(x)\ F(0, q) = 0. \tag{16.27}$$

These equations are to be interpreted as conditions which restrict the choice of the initial Schrödinger function $F(0, q)$: only such functions $F(0, q)$ are admissable which can be made to vanish by operating on them with $\pi_4(x)$ and $\nabla\cdot\pi(x)$ for any x. If we now apply the time-dependent operator (16.25) to $F(0, q)$, the first two terms of the series will give zero from (16.26). The same is also true for all higher terms; because according to (16.17):

$$\ddot{\chi} = c^2\nabla^2\chi, \qquad \dddot{\chi} = c^2\nabla^2\dot{\chi},$$
$$\ddddot{\chi} = c^4\nabla^4\chi, \qquad \text{etc.},$$

while on the other hand from (16.26):

$$\nabla^2\chi(x)\ F(0, q) = 0, \qquad \nabla^2\dot{\chi}(x)\ F(0, q) = 0,$$
$$\nabla^4\chi(x)\ F(0, q) = 0, \qquad \text{etc.}$$

In this way the equations (16.24 and .22) are a consequence of the initial conditions (16.26 or 16.27). On account of (16.22) we have further:

$$(16.28)\quad 0 = \left(H + \frac{h}{i}\frac{\partial}{\partial t}\right)\pi_4(x)\ F(t, q) = \left(H\,\pi_4(x) + \pi_4(x)\,\frac{h}{i}\frac{\partial}{\partial t}\right)F(t, q)$$
$$= [H\,\pi_4(x) - \pi_4(x)\,H]\,F(t, q) = \frac{h}{i}\,\dot{\pi}_4(x)\ F(t, q),$$

hence according to (16.15):

$$(16.29)\qquad \nabla\cdot\pi(x)\ F(t, q) = 0.$$

The operators $\pi_4(x)$ and $\nabla\cdot\pi(x)$ give zero when applied to the Schrödinger function for all times, if this was the case for $t = 0$.

The result of our discussion is that the problem of the Maxwell vacuum field can be defined by the Hamiltonian (16.13), if we add to the Schrödinger function the initial conditions (16.26 or 16.27) or the subsidiary conditions (16.22) which are equivalent to them. We shall verify that under these conditions Maxwell's equations still hold as operator equations. The canonical field equations for $\dot{\pi}$ (16.15) in conjunction with the subsidiary condition (16.22) yields:

$$(\dot{\pi} + \nabla\times(\nabla\times\psi))\ F(t, q) = 0;$$

Since according to (16.1, 16.11):

$$\mathfrak{E} = -c\,\pi, \qquad \mathfrak{H} = \nabla\times\psi$$

we can write for this also:

$$\left(-\frac{1}{c}\dot{\mathfrak{E}} + \nabla\times\mathfrak{H}\right)\ F(t, q) = 0.$$

(16.29) signifies:

$$\nabla \cdot \mathfrak{E} \, F(t, q) = 0.$$

On the other hand it follows from (16.12), that the curl of the operator $(\dot{\psi} - c^2 \pi)$ vanishes:

$$\left(\frac{1}{c} \dot{\mathfrak{H}} + \nabla \times \mathfrak{E}\right) = 0,$$

while $\mathfrak{H} = \nabla \times \psi$ has no sources by definition:

$$\nabla \cdot \mathfrak{H} = 0.$$

Thus we have derived all of Maxwell's equations. If we calculate further the expectation values of the field strengths:[1]

$$\overline{\mathfrak{E}}(x, t) = \int dq \, F^*(t, q) \; \mathfrak{E}(x) \cdot F(t, q),$$

$$\overline{\mathfrak{H}}(x, t) = \int dq \, F^*(t, q) \; \mathfrak{H}(x) \cdot F(t, q),$$

then these, too, satisfy as space-time functions the Maxwell equations:

$$-\frac{1}{c} \frac{\partial \overline{\mathfrak{E}}}{\partial t} + \nabla \times \overline{\mathfrak{H}} = 0, \qquad \nabla \cdot \overline{\mathfrak{E}} = 0,$$

$$\frac{1}{c} \frac{\partial \overline{\mathfrak{H}}}{\partial t} + \nabla \times \overline{\mathfrak{E}} = 0, \qquad \nabla \cdot \overline{\mathfrak{H}} = 0.$$

The propagation of a wave field $\overline{\mathfrak{E}}(x, t)$, $\overline{\mathfrak{H}}(x, t)$ *in vacuo* takes place according to the laws of the electromagnetic theory of light.

The subsidiary condition (16.22) permits a simplification of the Hamiltonian operator H. Adding to H (16.13) the space divergence:

$$-i c \, \nabla \cdot (\pi \, \psi_4),$$

one obtains the equivalent Hamiltonian operator:

$$\frac{1}{2} c^2 \left(|\pi|^2 + \pi_4^2\right) + \frac{1}{2} |\nabla \times \psi|^2 - i c \left\{\psi_4 \; \nabla \cdot \pi + (\nabla \cdot \psi) \pi_4\right\}.$$

Applied to the Schrödinger function $F(t, q)$ the terms which contain the factors π_4 or $\nabla \cdot \pi$ are zero, according to (16.22,.29). Hence H in the Schrödinger equa-

[1] The Schrödinger function shall be normalized to unity:

$$\int dq \, F^*(t, q) \, F(t, q) = 1.$$

tion is replaceable by $\bar{H}_0 = \int dx\, H_0$, where:

$$H_0 = \frac{1}{2}\{c^2 |\pi|^2 + |\nabla\times\psi|^2\} = \frac{1}{2}\{|\mathfrak{E}|^2 + |\mathfrak{H}|^2\}. \tag{16.30}$$

H_0 is identical with the known energy density of the electromagnetic field,[1] i.e. $H_0 = -T_{44}$, if the electromagnetic energy-momentum tensor is defined as usual by:

$$T_{\mu\nu} = \frac{1}{2}\sum_{\varrho}(f_{\mu\varrho} f_{\nu\varrho} + f_{\nu\varrho} f_{\mu\varrho}) - \delta_{\mu\nu}\frac{1}{4}\sum_{\varrho\sigma} f^2_{\varrho\sigma} \tag{16.31}$$

As in the case of the vector meson field (§12) $T_{\mu\nu}$ is different from the canonical energy-momentum tensor (2.8) which is not symmetrical. For the expectation values of the $T_{\mu\nu}$ the conservation equations hold:

$$\sum_{\mu}\frac{\partial \bar{T}_{\mu\nu}}{\partial x_\mu} = 0$$

as a consequence of the Maxwell equations. We forego the proof since it follows exactly the pattern of the corresponding classical calculation.

We shall now again carry out the transition to momentum space with the Fourier series (5.1):

$$\psi_\nu(x) = V^{-1/2}\sum_k q_{\nu,k}\, e^{ikx}, \qquad \pi_\nu(x) = V^{-1/2}\sum_k p_{\nu,k}\, e^{-ikx}. \tag{16.32}$$

Since ψ_j, π_j $(j = 1, 2, 3)$ shall be real fields, that is Hermitian operators, one must require according to (5.2):

$$q_{j,-k} = \overset{*}{q}_{j,k}, \qquad p_{j,-k} = \overset{*}{p}_{j,k}; \tag{16.33}$$

On the other hand, the quantities ψ_4, π_4, are not Hermitian operators but $i\,\psi_4$, $i\,\pi_4$ are Hermitian, since we use the imaginary time coordinate x_4.

$$q_{4,-k} = -\overset{*}{q}_{4,k}, \qquad p_{4,-k} = -\overset{*}{p}_{4,k}. \tag{16.34}$$

For the operators $q_{\nu,k}$, $p_{\nu,k}$ the canonical commutation rules (5.4) hold:

$$[q_{\nu,k}, q_{\nu',k'}] = [p_{\nu,k}, p_{\nu',k'}] = 0, \quad [p_{\nu,k}, q_{\nu',k'}] = \frac{h}{i}\,\delta_{\nu\nu'}\,\delta_{kk'}. \tag{16.35}$$

If we now consider the Schrödinger function F as function of the variables

[1] The field strengths $\mathfrak{E}$ and $\mathfrak{H}$ are measured here in Heaviside units. If one uses ordinary units all energy quantities must be divided by 4π.

$q_{\nu,k}$, the conjugate momenta $p_{\nu,k}$ must be interpreted as differential operators:

$$p_{\nu,k} = \frac{h}{i} \frac{\partial}{\partial q_{\nu,k}};$$

because it is known that the commutation rules (16.35) are then satisfied, and this is in agreement with the Hermitian conditions (16.33, 16.34). We consider first the subsidiary condition (16.22), which means according to (16.32):

$$\sum_k e^{-ikx} \, p_{4,k} F(t, q) = 0.$$

This shall be true for all positions x, consequently:

$$p_{4,k} F(t, q) \sim \frac{\partial}{\partial q_{4,k}} F(t, q) = 0 \tag{16.36}$$

for all k, i.e., $F(t, q)$ must be independent of the $q_{4,k}$. Another simple result follows from the equation (16.29), derived from the subsidiary condition (16.22). Obviously:

$$\nabla \cdot \pi(x) \sim \sum_k e^{-ikx} \sum_j k_j \, p_{j,k} = \sum_k e^{-ikx} (k \cdot p_k),$$

where p_k, as in §12, represents the 3-vector with the components $p_{1,k}$, $p_{2,k}$, $p_{3,k}$; (16.29) thus means:

$$(k \cdot p_k) F(t, q) = 0.$$

If we decompose the vectors q_k, p_k with the help of the coordinate axis $\mathfrak{e}_k^{(r)}$ (12.29, 12.30) into longitudinal and transverse components:

$$q_k = \sum_r \mathfrak{e}_k^{(r)} q_k^{(r)}, \qquad p_k = \sum_r \mathfrak{e}_k^{(r)} p_k^{(r)} \qquad (\mathfrak{e}_k^{(1)} \parallel k), \tag{16.37}$$

it follows that:

$$p_k^{(1)} F(t, q) \sim \frac{\partial}{\partial q_k^{(1)}} F(t, q) = 0, \tag{16.38}$$

i.e., F does not depend on the longitudinal components[1] either. Only the transverse components $q_k^{(2)}$, $q_k^{(3)}$ remain as essential variables.

We know already that the Hamiltonian operator $\boldsymbol{H}$ can be replaced in the

[1] It is well known that in classical theory $\nabla \cdot \pi = 0$ or $\nabla \cdot \mathfrak{E} = 0$ also entails the transverse character of the light waves.

Schrödinger equation by $\boldsymbol{H}_0$ (16.30). By substituting the Fourier series (16.32), we get:

$$H_0 = \int_V dx\, \boldsymbol{H}_0 = \frac{1}{2} \sum_k \{c^2 (p_k^* \cdot p_k) + ([k \times q_k^*] \cdot [k \times q_k])\},$$

or, using the component representation (16.37):

$$(16.39) \quad \begin{cases} H_0 = H^{\text{long}} + H^{\text{tr}}, \\ H^{\text{long}} = \dfrac{c^2}{2} \displaystyle\sum_k p_k^{(1)*} p_k^{(1)}, \quad H^{\text{tr}} = \dfrac{1}{2} \sum_k \sum_{r=2,3} \{c^2 p_k^{(r)*} p_k^{(r)} + k^2 q_k^{(r)*} q_k^{(r)}\}. \end{cases}$$

Since according to (16.38) $H^{\text{long}} F(t, q) = 0$, the Schrödinger equation reduces to:

$$(16.40) \qquad \left(\frac{h}{i} \frac{\partial}{\partial t} + H^{\text{tr}}\right) F(t, q) = 0.$$

After having thus eliminated the variables $q_{4,k}$ and $q_k^{(1)}$, we shall now change again to the matrix representation in order to determine the eigen values of H^{tr}; the solution can be obtained now exactly as in the case of the real meson field (§6). Analogous to (6.20) we substitute in (16.37):

$$(16.41) \quad q_k^{(r)} = \sqrt{\frac{h c}{2|k|}} (a_k^{(r)} + a_{-k}^{(r)*}), \qquad p_k^{(r)} = \sqrt{\frac{h|k|}{2c}}\, i\, (a_k^{(r)*} - a_{-k}^{(r)}).$$

With this the Hermitian conditions (16.33) are satisfied, provided that the coordinate axes $\mathfrak{e}_k^{(r)}$ and $\mathfrak{e}_{-k}^{(r)}$ belonging to two vectors k and $-k$ are selected parallel to each other and in the same direction.

$$(16.42) \qquad \mathfrak{e}_k^{(r)} = + \mathfrak{e}_{-k}^{(r)}.$$

[This results in a change of signs in the equation (12.30) for half of the k values, but these signs have no bearing on the following discussions.] $a_k^{(r)}$, $a_k^{(r)*}$ are matrices with respect to integers $N_k^{(r)}$ ($\geqq 0$), with the commutators:

$$(16.43) \quad [a_k^{(r)}, a_{k'}^{(r')}] = [a_k^{(r)*}, a_{k'}^{(r')*}] = 0, \quad [a_k^{(r)}, a_{k'}^{(r')*}] = \delta_{rr'}\, \delta_{kk'};$$

In this way we get:

$$[q_k^{(r)}, q_{k'}^{(r')}] = [p_k^{(r)}, p_{k'}^{(r')}] = 0, \quad [p_k^{(r)}, q_{k'}^{(r')}] = \frac{h}{i}\, \delta_{rr'}\, \delta_{kk'},$$

which corresponds with (16.37) to the commutation rules (16.35). The matrices $(a_k^{(r)*}\, a_k^{(r)})$ are diagonal and have the eigen values

$$(a_k^{(r)*}\, a_k^{(r)})_N = N_k^{(r)}. \tag{16.44}$$

For H^{tr} (16.39) we obtain with (16.41) the diagonal matrix:

$$H^{tr} = \frac{h\,c}{2} \sum_k \sum_{r=2,3} |k|\, (a_k^{(r)*}\, a_k^{(r)} + a_k^{(r)}\, a_k^{(r)*})$$

with the eigen values:

$$H_N = \sum_k h\,c\,|k|\,\,(N_k^{(2)} + N_k^{(3)} + 1). \tag{16.45}$$

If one considers the Schrödinger function F, in accordance with the previously used matrix representation, as function of the variables $N_k^{(r)}$ the general solution of the Schrödinger equation (16.40) is:

$$F\,(t, N) = F\,(0, N)\, e^{-\frac{i\,t}{h} H_N};$$

$F\,(0, N)$ is the "probability amplitude" of the stationary state of the Maxwell field, characterized by the quantum numbers $N_k^{(r)}$. Apart from the zero point energy $h\,c \sum_k |k|$ cf. (6.23), the energy eigen values (16.45) correspond to the corpuscular interpretation of light: $N_k^{(2)} + N_k^{(3)}$ is the number of light quanta of the momentum $h\,k$ and the energy $h\,c|k|$, and the upper indices ($r = 2, 3$) obviously distinguish between the two transverse light polarizations.

In order to confirm this interpretation we shall discuss further the field momentum, which in the classical theory is given by a density $1/c\ \mathfrak{E} \times \mathfrak{H}$. The symmetrized operator is in accordance with (16.31) ($\mathsf{G}_j = T_{4j}/i\,c$) given by:

$$\begin{aligned} \mathsf{G} &= \frac{1}{2\,c} \{\, \mathfrak{E} \times \mathfrak{H} - \mathfrak{H} \times \mathfrak{E} \,\} \\ &= -\frac{1}{2} \{[\pi \times (\nabla \times \psi) - (\nabla \times \psi) \times \pi]\}. \end{aligned} \tag{16.46}$$

We substitute here again the Fourier series (16.32) and calculate the total momentum $G = \int dx\, \mathsf{G}$. To abbreviate the computation we suppress the terms which contain a factor $p_k^{(1)}$ by assuming that the operator G is always applied to the Schrödinger function F. It follows then:

$$G = -\frac{i}{2} \sum_k k \sum_{r=2,3} (p_k^{(r)}\, q_k^{(r)} + q_k^{(r)}\, p_k^{(r)}) \tag{16.47}$$

and with (16.41):[1]

$$G = \frac{h}{2} \sum_k k \sum_{r=2,3} (a_k^{(r)*} a_k^{(r)} + a_k^{(r)} a_k^{(r)*}).$$

This diagonal matrix has the eigen values:

$$G_N = \sum_k h\, k \cdot (N_k^{(2)} + N_k^{(3)}), \tag{16.48}$$

which—as we expected—represent the resulting momenta of all photons present.

Each state $(N_k^{(2)}, N_k^{(3)})$ must be counted as a single state, in accordance with the Bose postulate for enumerating the states of a "gas of light quanta."

The fact that the number of light quanta of given momentum can be represented as sum of two independent integers indicates that the light quantum has a spin with two possibilities of orientation. In order to study its properties, we decompose the angular momentum:

$$M = \int dx\, x \times G = -\frac{1}{2} \int dx \left\{ x \times [\pi \times (\nabla \times \psi) - (\nabla \times \psi) \times \pi] \right\} \tag{16.49}$$

according to the pattern of the formulas (12.50 to 12.60) into two terms M^0 and M'. Omitting again terms containing the factor $\nabla \cdot \pi$ in view of (16.29, 16.38), we have:

$$\begin{cases} M = M^0 + M', \\ M^0_{jj'} = -\frac{1}{2} \int dx \sum_i \left\{ x_j \left(\pi_i \frac{\partial \psi_i}{\partial x_{j'}} + \frac{\partial \psi_i}{\partial x_{j'}} \pi_i \right) \right. \\ \qquad\qquad \left. - x_{j'} \left(\pi_i \frac{\partial \psi_i}{\partial x_j} + \frac{\partial \psi_i}{\partial x_j} \pi_i \right) \right\}, \\ M'_{jj'} = -\int dx \left\{ \pi_j \psi_{j'} - \pi_{j'} \psi_j \right\}. \end{cases} \tag{16.50}$$

We can again prove that the expectation value of M^0 is independent of the polarization of the light quanta; hence the spin must be contained in M'. The transition to momentum space yields:

$$M' = \sum_k M'_{(k)}, \qquad M'_{(k)} = -\, p_k \times q_k\,. \tag{16.51}$$

Here, as in §12, each term $M'_{(k)}$ of the sum can be discussed separately. Unlike

[1] Compare the analogous formulas (6.26, 6.27, 6.28).

the treatment in §12, it is impossible to discuss the problem in the rest system of the particles because there is no rest system, the velocity being $d\omega_k/d|k| = c$ in every coordinate system. We pick out an arbitrary term $M'_{(k)}$ of the sum (16.51) together with the term $M'_{(-k)}$ [in view of the fact that the operators q_k, p_k according to (16.41) operate on the light quantum numbers $N_k^{(r)}$ and $N_{-k}^{(r)}$ and consider the longitudinal component of the angular momentum $(M'_{(k)} + M'_{(-k)})$, i.e., its projection onto the propagation direction of the light quantum $\mathfrak{e}_k^{(1)} = \mathfrak{e}_{-k}^{(1)}$, cf. (16.42)]:

$$(M'_{(k)} + M'_{(-k)}) \cdot \mathfrak{e}_k^{(1)} = - \{p_k^{(2)} q_k^{(3)} - p_k^{(3)} q_k^{(2)} + p_{-k}^{(2)} q_{-k}^{(3)} - p_{-k}^{(3)} q_{-k}^{(2)}\}.$$

We obtain for it with the matrix representation (16.41):

$$(M'_{(k)} + M'_{(-k)}) \cdot \mathfrak{e}_k^{(1)} = h\, i \, \{(- a_k^{(2)*} a_k^{(3)} + a_k^{(3)*} a_k^{(2)}) + (- a_{-k}^{(2)*} a_{-k}^{(3)} + a_{-k}^{(3)*} a_{-k}^{(2)})\}.$$

Here the contributions of the photons with the momenta $+ h\, k$ and $- h\, k$ are separated. Each single contribution has the form (12.61) and hence can be made diagonal by canonical transformations of the type (12.62) (which correspond to the transition from linear to circular polarized light waves). According to (12.63) the result may be stated: the longitudinal component of the spin of the light quanta with momentum $h\, k$ has the eigen values:

$$h\, (N_{k+} - N_{k-}),$$

where N_{k+} and N_{k-} are two non-negative integers, the sum of which equals the total number of light quanta with the momentum $h\, k$:

$$N_{k+} + N_{k-} = N_k^{(2)} + N_k^{(3)}.$$

This result can be interpreted that the photon has the spin h (the photon is a corpuscle with "spin 1"), but only alignments parallel or antiparallel to the direction of propagation k are possible: N_{k+} and N_{k-} represent the number of spins oriented parallel or antiparallel, respectively.

§ 17. Interaction with Electrons

In the presence of electric charges or currents, the Maxwell equations read:

$$f_{\mu\nu} = \frac{\partial \varphi_\nu}{\partial x_\mu} - \frac{\partial \varphi_\mu}{\partial x_\nu}, \quad \sum_\mu \frac{\partial f_{\mu\nu}}{\partial x_\mu} = - \frac{1}{c} s_\nu, \tag{17.1}$$

where, as before, $s_4/i\,c = \rho$ stands for the charge density and s for the current density. In classical theory the s_ν are space-time functions, which satisfy the continuity equation:

$$\sum_\nu \frac{\partial s_\nu}{\partial x_\nu} = 0 \quad \text{or} \quad \frac{\partial \varrho}{\partial t} + \nabla \cdot s = 0 \tag{17.2}$$

If we deal in particular with point charges e_n which move along certain paths $x = x_n(t)$, the charge density at the respective positions will be singular as $e_n \delta\,(x - x_n)$. Similarly the current density which equals charge density multiplied with the velocity:

$$\varrho\,(x, t) = \sum_n e_n\, \delta\big(x - x_n\,(t)\big), \qquad s\,(x, t) = \sum_n e_n\, \dot{x}_n(t)\, \delta\big(x - x_n\,(t)\big); \tag{17.3}$$

the continuity equation is satisfied, since:

$$\frac{\partial \varrho}{\partial t} = -\sum_n e_n\, \Big(\dot{x}_n(t) \cdot \nabla\, \delta\big(x - x_n\,(t)\big)\Big) = -\nabla \cdot s.$$

We discuss first ρ and s as given space-time functions and for this case we generalize the canonical formalism, used in §16. Let:

$$L = L^0 + L', \qquad L' = \frac{1}{c} \sum_\nu s_\nu\, \psi_\nu, \tag{17.4}$$

where L^0 is the Lagrangian of the vacuum field (16.5). Instead of the field equation (16.6) we obtain in this case:

$$\Box\, \psi_\nu + \frac{1}{c}\, s_\nu = 0, \tag{17.5}$$

or

$$\sum_\mu \frac{\partial f_{\mu\nu}}{\partial x_\mu} + \frac{\partial \chi}{\partial x_\nu} + \frac{1}{c}\, s_\nu = 0, \tag{17.6}$$

where χ is again defined by (16.8):

$$\chi = \sum_\mu \frac{\partial \psi_\mu}{\partial x_\mu}.$$

From (17.5) follows in connection with the continuity equation (17.2):

$$\Box\, \chi = \sum_\mu \frac{\partial}{\partial x_\mu}\, \Box\, \psi_\mu = -\frac{1}{c} \sum_\mu \frac{\partial s_\mu}{\partial x_\mu} = 0.$$

From this we can deduce, as in the case of the vacuum field, that χ vanishes identically if χ and $\partial\chi/\partial t$ vanish for $t = 0$. In this case (17.6) goes over into Maxwell's equations (17.1). The defining equations (16.11) of the fields π_ν which are canonically conjugate to ψ_ν are not changed. The Hamiltonian $H^0 + H'$ corresponding to the Lagrangian (17.4) is given by:

$$H^0 = \frac{1}{2} c^2 \left(|\pi|^2 + \pi_4^2\right) + \frac{1}{2} |\nabla \times \psi|^2 + i\, c \left\{(\pi \cdot \nabla\, \psi_4) - \pi_4\, \nabla \cdot \psi\right\}, \tag{17.7}$$

and $H' = -L'$:

$$H' = -\frac{1}{c} \sum_\nu s_\nu\, \psi_\nu = -\frac{1}{c}\,(s \cdot \psi) - i\, \varrho\, \psi_4. \tag{17.8}$$

The quantization of the field can be carried through according to the canonical commutation rules (16.14).[1]

We shall now no longer make the assumption that the charges and currents are given, in order to consider also the influences of the electromagnetic field on the charge-carrying fields or particles. For this purpose we must express the latter also by a Lagrangian or a Hamiltonian. The Lagrangian of a complex scalar field Ψ for instance, can be written [(cf. (11.2)]:

$$-c^2 \left\{ \sum_\nu \left(\frac{\partial \Psi^*}{\partial x_\nu} + \frac{i\, \varepsilon}{c}\, \psi_\nu\, \Psi^* \right) \left(\frac{\partial \Psi}{\partial x_\nu} - \frac{i\, \varepsilon}{c}\, \psi_\nu\, \Psi \right) + \mu^2\, \Psi^*\, \Psi \right\};$$

where we have put ψ_ν for the electromagnetic potential. If one adds to this expression L^0 (16.5), one obtains a Lagrangian which describes the interaction of an electromagnetic and a scalar field. Indeed, we obtain the field equation for Ψ and Ψ^* in agreement with (11.1), and the equations for the ψ_ν change into (17.5 or 17.6), if the 4-current s_ν carried by the scalar field is defined according to (11.3).

More important than the interaction with hypothetical scalar particles is the one with electrons, which obeys the Dirac wave equation. We shall now discuss this particular problem. Since we shall not deal until later (Chapter V) with the quantization of the electron wave field, according to the Pauli exclusion principle, we must be content with a description of the electrons in configuration

[1] The subsidiary condition $\chi = 0$ must be replaced by the operator equation (16.22) as for the case of a vacuum field; see below. One can prove the Lorentz invariance of the quantization method, with a method similar to the one used in §7 (meson field coupled with a scalar density function η). We do not need to enter into this, since questions of invariance shall be discussed in more general connection in §18.

space (wave mechanical formalism without "second" quantization).[1]

According to Dirac the nth particle is described by a Hamiltonian operator:

$$H_{(n)} = m_n c^2 \beta_n + c\left(\alpha_n \cdot \left\{p_n - \frac{e_n}{c} \psi(x_n)\right\}\right) - e_n \cdot i \psi_4(x_n); \tag{17.9}$$

β_n and the three components of the vector α_n are the known Dirac matrices with respect to the spin coordinates of the nth electron; and p_n signifies the momentum vector, represented as a differential operator with regard to the space vector x_n:

$$p_n = \frac{h}{i} \nabla_n. \tag{17.10}$$

We construct the Hamiltonian of the total system (field plus electrons):

$$H = H^0 + \sum_n H_{(n)}, \tag{17.11}$$

where $H^0 = \int dx\, \mathsf{H}^0$ (17.7) corresponds to the vacuum field. Calling the contribution of the free electrons H^P:

$$H^P = \sum_n \{m_n c^2 \beta_n + c(\alpha_n \cdot p_n)\}, \tag{17.12}$$

we can write instead of (17.11):

$$H = H^0 + H^P + H', \tag{17.13}$$

where H' represents the interaction terms:

$$H' = -\sum_n e_n \{(\alpha_n \cdot \psi(x_n)) + i \psi_4(x_n)\}. \tag{17.14}$$

This expression for the interaction agrees, by the way, with (17.8), because if one substitutes there for the charge and current densities of the electrons:

$$\varrho(x) = \sum_n e_n \delta(x - x_n), \qquad s(x) = c \sum_n e_n \alpha_n \delta(x - x_n) \tag{17.15}$$

[cf. (17.3), $c\,\alpha_n$ corresponds to the velocity $\dot{x}_n$], $\int dx\, \mathsf{H}'$ reduces to H' (17.14). For the charge quantities (17.15) the conservation law is true in the form of the

[1] We treated the proton-neutron in the problem of interaction with meson fields (§§9 and 14) in the same way; we only simplified the problem there, by assuming that the proton-neutrons are infinitely heavy, so that translatory degrees of freedom could be disregarded.

operator equation:

$$(17.16)\qquad \dot{\varrho}(x) = \frac{i}{h}[H, \varrho(x)] = \frac{i}{h}[H^P, \varrho(x)]$$
$$= c\sum_n e_n\big(\alpha_n \cdot \nabla_n \delta(x - x_n)\big) = -\nabla \cdot s(x).$$

With (17.11 or 17.13) we obtain the Schrödinger equation:

$$(17.17)\qquad \left(\frac{h}{i}\frac{\partial}{\partial t} + H\right) F(t, q, q_1, q_2, \ldots) = 0;$$

the Schrödinger function F depends now not only on the field variables q [$= \psi_\nu(x)$, cf. §16], but also on the electron coordinates q_1, q_2, ... where q_n denotes the space and spin coordinates of the nth electron.[1]

We do not need to discuss the motion of the electrons in the field since this is described by the Schrödinger equation (17.17) in agreement with the known Dirac theory of the electron. We shall, therefore, return to a discussion of the field. On account of (17.13) [with $H' = \int dx\, \mathbf{H}'$, where $\mathbf{H}'$ has the form (17.8)] we obtain now instead of the vacuum equations (16.12, 16.15, 16.16, 16.17) the following field equations:

$$(17.18)\qquad \begin{cases} \dot{\psi} = c^2 \pi + i c \nabla \psi_4, & \dot{\psi}_4 = c^2 \pi_4 - i c \nabla \cdot \psi, \\ \dot{\pi} = -\nabla \times (\nabla \times \psi) - i c \nabla \pi_4 + \dfrac{s}{c}, & \dot{\pi}_4 = i(c \nabla \cdot \pi + \varrho); \end{cases}$$

$$(17.19)\qquad \begin{cases} \ddot{\psi} = c^2 \nabla^2 \psi + c s, & \ddot{\psi}_4 = c^2 \nabla^2 \psi_4 + c s_4, \\ \ddot{\pi} = c^2 \nabla^2 \pi + \dfrac{\dot{s}}{c} + c \nabla \varrho, & \ddot{\pi}_4 = c^2 \nabla^2 \pi_4. \end{cases}$$

These equations correspond to the field equations (17.5, 17.6) which were too general. To obtain the Maxwell equations we still require the counterpart to the classical subsidiary condition $\chi = 0$, respectively to the initial condition, "$\chi = 0$, $\partial\chi/\partial t = 0$ for $t = 0$." According to the second equation (17.18):

$$\chi \equiv \frac{1}{ic}\dot{\psi}_4 + \nabla \cdot \psi = -i c \pi_4.$$

[1] As usual in Dirac's wave mechanics we can assume the spin coordinates written as indices on F; the operators α_n, β_n operate then on the nth spin index according to the well-known rules of matrix multiplications, while they are unit matrices with regard to the other spin indices ($n' \neq n$). We shall discuss in §18 the question of the relativistic and gauge invariance of the formalism.

As in the case for vacuum, we must therefore impose on the Schrödinger function F the following initial conditions:

$$(17.20) \qquad \pi_4(x)\cdot F(0, q, q_1, \ldots) = 0, \qquad \dot{\pi}_4(x)\cdot F(0, q, q_1, \ldots) = 0.$$

Since according to (17.19):

$$\ddot{\pi}_4 = c^2 \nabla^2 \pi_4, \quad \dddot{\pi}_4 = c^2 \nabla^2 \dot{\pi}_4, \quad \ddddot{\pi}_4 = c^4 \nabla^4 \pi_4 \text{ etc.},$$

it follows as in §16, that all operators $\ddot{\pi}_4(x)$, $\dddot{\pi}_4(x)$, $\ddddot{\pi}_4(x)$, etc., applied to $F(0, q, \ldots)$ become zero; and with the help of the formulas (16.23, .25, .21) (where $\chi = -ic\pi_4$ and $H = H^0 + H^P + H'$) one can deduce the validity of the equations (16.24, 16.22, 16.28), as in the case of the vacuum:

$$(17.21) \qquad \pi_4(x)\ F(t, q, q_1, \ldots) = 0, \qquad \dot{\pi}_4(x)\ F(t, q, q_1, \ldots) = 0.$$

These subsidiary conditions for F are thus permanently fulfilled, provided they hold for $t = 0$. The second subsidiary condition (17.21) implies according to (17.18):

$$(17.22) \qquad (c\nabla\cdot\pi + \varrho)\ F(t, q, q_1, \ldots) = 0$$

or on account of $\pi_j = f_{j4}/ic = -\mathfrak{E}_j/c$:

$$(\nabla\cdot\mathfrak{E} - \varrho)\ F(t, q, q_1, \ldots) = 0,$$

which corresponds to the classical equation $\nabla\cdot\mathfrak{E} = \rho$. Furthermore the third equation (17.18) yields in connection with (17.21):

$$(\dot{\mathfrak{E}} - c\ \nabla\times\mathfrak{H} + s)\ F(t, q, q_1, \ldots) = 0,$$

and the rest of the Maxwell equations are valid without any change as compared with the case for the vacuum. It results in particular from this that the expectation values $\overline{\mathfrak{E}}(x, t)$, $\overline{\mathfrak{H}}(x, t)$, $\overline{s}(x, t)$, $\overline{\rho}(x, t)$ satisfy Maxwell's equations:

$$-\frac{1}{c}\frac{\partial\overline{\mathfrak{E}}}{\partial t} + \nabla\times\overline{\mathfrak{H}} = \frac{1}{c}\overline{s}, \qquad \nabla\cdot\overline{\mathfrak{E}} = \overline{\varrho},$$

$$\frac{1}{c}\frac{\partial\overline{\mathfrak{H}}}{\partial t} + \nabla\times\overline{\mathfrak{E}} = 0, \qquad \nabla\cdot\overline{\mathfrak{H}} = 0.$$

These equations represent the quantum theoretical interpretation of the classical field equations. The classical field functions correspond to the expectation values of the field operators. Similarly the conservation laws for the energy-momentum tensor (16.31) can be derived, as in the classical theory, in the form:

$$\sum_\mu \frac{\partial \overline{T}_{\mu\nu}}{\partial x_\mu} = \frac{1}{c} \overline{\sum_\mu s_\mu f_{\mu\nu}}.$$

The Hamiltonian operator in the Schrödinger equation can again be simplified as a consequence of the subsidiary conditions. In the first place, the terms containing π_4 can be dropped in H^0 (17.7). Furthermore the term:

$$i\,c\,(\pi \cdot \nabla\,\psi_4)$$

can be changed into $-\,i\,c\,\psi_4 \nabla \cdot \pi$ by adding the divergence $-\,i\,c\,\nabla \cdot (\pi\,\psi_4)$ and this is, if applied to F, equivalent to $i\,\rho\,\psi_4$ according to (17.22). This term, however, compensates the term $-i\rho\psi_4$ in H' (17.8). As a result, $H^0 + H'$ in the Schrödinger equation can be replaced by $H_0 + H_1$, where:

$$H_0 = \frac{1}{2}\left\{c^2 |\pi|^2 + |\nabla \times \psi|^2\right\} \tag{17.23}$$

[cf. (16.30)] is the electromagnetic energy density and where:

$$H_1 = -\frac{1}{c}(s \cdot \psi), \qquad \int dx\, H_1 = -\sum_n e_n \big(\alpha_n \cdot \psi\,(x_n)\big). \tag{17.24}$$

As new Schrödinger equation, we obtain:

$$\left(\frac{h}{i}\frac{\partial}{\partial t} + \int dx\,(H_0 + H_1) + H^P\right) F\,(t, q, q_1, \ldots) = 0. \tag{17.25}$$

Further calculations are conveniently handled in momentum space. Again we use the Fourier series (16.37) with coefficients $q_{\nu,\,k}$, $p_{\nu,\,k}$ subjected to the Hermitian conditions (16.33, 16.34) and the canonical commutation rules (16.35). The subsidiary condition $\pi_4\,(x)F = 0$ then states according to (16.36) that the Schrödinger function is independent of the $q_{4,k}$:

$$\frac{\partial}{\partial q_{4,\,k}} F\,(t, q, \ldots) = 0. \tag{17.26}$$

We substitute (16.32) into the second subsidiary condition (17.22):

$$\nabla \cdot \pi = -\,i\,V^{-1/2} \sum_k (k \cdot p_k)\, e^{-ikx}$$

and we also expand the charge density ρ in a spatial Fourier series:

$$\varrho = V^{-1/2} \sum_k \varrho_k e^{-ikx}; \tag{17.27}$$

on account of the reality of ρ, we have here:

(17.28) $$\varrho_{-k} = \overset{*}{\varrho}_k.$$

In order that (17.22) be satisfied for all x, we must have:

$$\left\{- i\,(k\cdot p_k) + \frac{\varrho_k}{c}\right\} F\,(t, q, \ldots) = 0,$$

and this must be valid for all k values. Considering F again as function of the longitudinal components $q_k^{(1)}$ and transverse components $q_k^{(2)}$, $q_k^{(3)}$; cf. (16.37):

(17.29) $$(k\cdot p_k) = |k|\,p_k^{(1)} = |k|\,\frac{h}{i}\,\frac{\partial}{\partial q_k^{(1)}};$$

we obtain the differential equations:

$$\left\{\frac{\partial}{\partial q_k^{(1)}} - \frac{1}{c\,h}\,\frac{\varrho_k}{|k|}\right\} F\,(t, q, \ldots) = 0.$$

The solution is:

(17.30) $$F\,(t, q, \ldots) = e^{\frac{1}{c\,h}\sum_k \frac{\varrho_k}{|k|}\,q_k^{(1)}} \cdot F^{\mathrm{tr}}\,(t, q, \ldots),$$

where F^{tr} is independent of the $q_k^{(1)}$:

(17.31) $$\frac{\partial}{\partial q_k^{(1)}}\,F^{\mathrm{tr}}\,(t, q, \ldots) = 0.$$

In (17.29) $q_k^{(1)}$ and $p_k^{(1)}$ signify the components of q_k and p_k in the direction of $+\,k$:

$$q_k^{(1)} = q_k\cdot \mathfrak{e}_k^{(1)}\,,\quad p_k^{(1)} = p_k\cdot \mathfrak{e}_k^{(1)}\,,\quad \text{wo}\quad \mathfrak{e}_k^{(1)} = +\,\frac{k}{|k|}.$$

Thus:

(17.32) $$\mathfrak{e}_{-k}^{(1)} = -\,\mathfrak{e}_k^{(1)},$$ [1]

and the Hermitian condition (16.33) implies that:

(17.33) $$q_{-k}^{(1)} = -\,q_k^{(1)*},\qquad p_{-k}^{(1)} = -\,p_k^{(1)*}.$$

From this and from (17.28) it is evident that the sum $\sum_k \rho_k q_k^{(1)}/|k|$, appearing in the exponent of (17.30), is purely imaginary.

[1] Since the coordinate vectors $\mathfrak{e}_k^{(r)}$ are not necessarily a right-handed system it is possible to orient the transverse axes $\mathfrak{e}_k^{(r)}$ and $\mathfrak{e}_{-k}^{(r)}$ parallel. $\mathfrak{e}_{-k}^{(2)} = \mathfrak{e}_k^{(2)}$, $\mathfrak{e}_{-k}^{(3)} = \mathfrak{e}_k^{(3)}$. This is an advantage for the discussion of the transverse field, since the formulas (16.41 ff.) can be used.

We shall now substitute (17.30) into the Schrödinger equation (17.25). For the field energy $\int dx\, H_0$ we get as in the case of the vacuum [cf. (16.39)]:

$$(17.34)\quad \begin{cases} \int dx\, H_0 = H^{\text{long}} + H^{\text{tr}}, \\ H^{\text{long}} = \frac{c^2}{2}\sum_k p_k^{(1)*} p_k^{(1)} = -\frac{c^2}{2}\sum_k p_{-k}^{(1)} p_k^{(1)} = \frac{c^2 h^2}{2}\sum_k \frac{\partial^2}{\partial q_{-k}^{(1)}\, \partial q_k^{(1)}}, \\ H^{\text{tr}} = \frac{1}{2}\sum_k \sum_{r=2,3} \{c^2 p_k^{(r)*} p_k^{(r)} + k^2 q_k^{(r)*} q_k^{(r)}\}. \end{cases}$$

If we first apply the operator H^{long} to F, it follows with the help of (17.30, 17.31):

$$(17.35)\qquad H^{\text{long}} \cdot F(t, q, \ldots) = H^C \cdot F(t, q, \ldots),$$

where:

$$(17.36)\qquad H^C = \frac{1}{2}\sum_k \frac{1}{|k|^2}\, \varrho_{-k}\, \varrho_k$$

We claim: H^C is simply the Coulomb energy of the charge distribution $\rho(x)$ (17.27). For it is known that this energy equals $\frac{1}{2}\int dx\, \rho(x)\, \Phi(x)$, where Φ represents the Coulomb potential of the charges ρ, which is determined by the Poisson differential equation $\nabla^2 \Phi = -\rho$:

$$(17.37)\qquad \Phi(x) = V^{-1/2}\sum_k \frac{\varrho_k}{|k|^2}\, e^{-ikx}$$

[cf. (17.27)]. From this it follows that:

$$(17.38)\qquad \frac{1}{2}\int_{(V)} dx\, \varrho(x)\, \Phi(x) = \frac{1}{2}\sum_k \varrho_{-k} \frac{\varrho_k}{|k|^2} = H^C,$$

as has been stated. This holds also if the charge density degenerates into the singular density function (17.15). In this case (for the limit $V = \infty$):

$$\Phi(x) = \frac{1}{4\pi}\sum_n \frac{e_n}{|x - x_n|},\text{[1]}$$

and H^C becomes equal to the electrostatic energy of the point charges:

$$(17.39)\qquad H^C = \frac{1}{8\pi}\sum_{nn'} \frac{e_n\, e_{n'}}{|x_n - x_{n'}|}.$$

[1] The numerical factor $1/4\pi$ comes from the choice of the Heaviside units for the charges.

It should be noticed that the electrical self-energies of the particles ($n = n'$), which become infinite in the limiting case of point charges, appear here besides the Coulomb interaction energies ($n \neq n'$). If one attributes to the electron a finite extension, as in the classical Lorentz theory, the self-energies are, of course, finite. This procedure, however, compels us to relinquish the relativistic invariance, as was discussed in detail in §7 for the analogous problem of the meson field theory. We shall discuss in §19 another possibility to make the electrical self-energy finite or rather to eliminate it.

With (17.34, 17.35) the Schrödinger equation (17.25) can be written:

$$\left(\frac{h}{i}\frac{\partial}{\partial t} + H^{tr} + H^{C} + \int dx\, H_1 + H^{P}\right) F(t, q, \ldots) = 0,$$

or with (17.22, 17.24):

$$\left(\frac{h}{i}\frac{\partial}{\partial t} + H^{tr} + H^{C} + \sum_n \left\{m_n c^2 \beta_n + c\left(\alpha_n \cdot \left\{p_n - \frac{e_n}{c}\psi(x_n)\right\}\right)\right\}\right) F(t, q, \ldots) = 0. \tag{17.40}$$

We shall now move the exponential factor:

$$S = e^{\frac{1}{hc}\sum_k \frac{\varrho_k}{|k|} q_k^{(1)}} \tag{17.41}$$

contained in F (17.30) to the left of the Hamilton operator in (17.40), in order to obtain a Schrödinger equation for F^{tr} alone. S commutes with the operators $\partial/\partial t$, H^{tr} (17.34) and H^{C} (17.39), but not with the differential operators p_n (17.10), since S depends through the ρ_k on the electron coordinates x_n. According to (17.27 and .15):

$$\varrho_k = V^{-1/2}\int dx\, \varrho(x)\, e^{ikx} = V^{-1/2}\sum_n e_n\, e^{ikx_n}. \tag{17.42}$$

If one substitutes this into:

$$[p_n, S] = \frac{h}{i}\nabla_n S = \frac{1}{ci}\sum_k \frac{1}{|k|} q_k^{(1)} (\nabla_n \varrho_k)\cdot S,$$

it follows:

$$[p_n, S] = \frac{e_n}{c} V^{-1/2}\sum_k \frac{k}{|k|} q_k^{(1)} e^{ikx_n}\cdot S. \tag{17.43}$$

The sum over k has a simple meaning: let us separate the vector potential:

$$\psi(x) = V^{-1/2} \sum_k q_k e^{ikx} = V^{-1/2} \sum_k \left(\sum_r e_k^{(r)} q_k^{(r)} \right) e^{ikx}$$

into longitudinal and transverse waves:

$$(17.44) \quad \begin{cases} \psi = \psi^{\text{long}} + \psi^{\text{tr}}, \\ \psi^{\text{long}}(x) = V^{-1/2} \sum_k e_k^{(1)} q_k^{(1)} e^{ikx} = V^{-1/2} \sum_k \frac{k}{|k|} q_k^{(1)} e^{ikx}, \\ \psi^{\text{tr}}(x) = V^{-1/2} \sum_k (e_k^{(2)} q_k^{(2)} + e_k^{(3)} q_k^{(3)}) e^{ikx}, \end{cases}$$

then ψ^{long} appears in (17.43), taken at the point x_n:

$$[p_n, S] = \frac{e_n}{c} \psi^{\text{long}}(x_n) \; S.$$

It follows from this:

$$(17.45) \qquad \left\{ p_n - \frac{e_n}{c} \psi(x_n) \right\} S = S \left\{ p_n - \frac{e_n}{c} \psi^{\text{tr}}(x_n) \right\};$$

i.e., moving S towards the left side changes ψ into ψ^{tr}; ψ^{long} is eliminated. Hence if one sets in the Schrödinger equation (17.40) according to (17.30, 17.41): $F = SF^{\text{tr}}$ and if one moves S to the left, simply all terms ψ are changed into ψ^{tr}. After multiplication with S^{-1} [according to (17.28, 17.33): $SS^* = SS^* = 1$, hence $S^{-1} = S^*$], it follows finally:

$$(17.46) \quad \left(\frac{h}{i} \frac{\partial}{\partial t} + H^{\text{tr}} + H^C + H^P + H^W \right) F^{\text{tr}}(t, q, q_1, q_2, \ldots) = 0,$$

where H^P again denotes the Hamilton operator of the free particles (17.12), and H^W represents the following reduced interaction operator:

$$(17.47) \qquad H^W = -\frac{1}{c} \int dx \, (s \cdot \psi^{\text{tr}}) = -\sum_n e_n \left(\alpha_n \cdot \psi^{\text{tr}}(x_n) \right).$$

The longitudinal field quantities $q_k^{(1)}$, $p_k^{(1)}$ enter neither in H^{tr} nor in H^W, so that (17.46) is compatible with (17.31).

The subsidiary conditions (17.21, 17.22) have thus enabled us to reduce the problem of the interaction between electromagnetic field and electrons to the problem (17.46), where—besides the electron coordinates—only the coordinates $q_k^{(2)}$, $q_k^{(3)}$ of the transverse light waves appear. The Hamilton operator of this reduced problem contains: the energy of transverse light waves (H^{tr}), the kinetic and Coulomb energy of the electrons ($H^P + H^C$) and the term H^W, which couples electrons and light. One understands easily that the essential part of this result remains if, instead of the Dirac electrons, charged particles with different properties are introduced. If one describes for instance these particles by the Schrödinger (unrelativistic) wave equation, so that instead of (17.9), one has:

$$H_{(n)} = \frac{1}{2\,m_n}\left\{p_n - \frac{e_n}{c}\,\psi\,(x_n)\right\}^2 - e_n\,i\,\psi_4\,(x_n)$$

then the elimination of the $q_{4,k}$ and $q_k^{(1)}$ leads again to a Schrödinger equation of the type (17.46), where H^W is derived from H', by replacing ψ_4 by 0 and ψ by ψ^{tr}:

$$H^W = \sum_n \left\{-\frac{e_n}{m_n\,c}\left(\psi^{tr}\,(x_n)\cdot p_n\right) + \frac{e_n^2}{2\,m_n\,c^2}\left(\psi^{tr}\,(x_n)\right)^2\right\}. \tag{17.48}$$

The situation is similar in the case of charged particles represented by quantized waves, for instance, in the case of the scalar complex meson field (see above).

The reduced Schrödinger equation (17.46) is essentially the one on which Dirac[1] based his well-known theory of radiative transitions in 1927. Dirac divided the electromagnetic interactions into static (Coulomb) potentials, which he included in the Hamiltonian of the atomic systems in question, and into interactions between atoms and transverse light waves which were quantized. This division was of course not satisfactory from the point of view of the Maxwell theory which considers the static field and the light waves as a uniform entity. Only after Heisenberg, Pauli, and Fermi[2] had incorporated Dirac's theory into their general theory of quantum electrodynamics, was it possible to return to the unitary field concept and to establish the connection with the classical electrodynamics of macrophysics. Furthermore, the invariance of the formalism under Lorentz transformations can only be proved in the general quantum electrodynamics, since these transformations transform the static and the wave parts of the field into each other. In the applications of the theory to special problems, the reduced problem (17.46), i.e., the Dirac radiation theory, is most convenient. We do not intend to enter here into the many important

[1] *Proc. Roy. Soc. London 114*, 243a, 710, 1927.

[2] Quotations on p. 112.

results of this theory,[1] but rather restrict ourselves to discussing the most immediate applications. We shall give a simple example.

If we consider in (17.46) the coupling term H^W as a small perturbation, then the variables of field and particle in the unperturbed problem ($H^W \to 0$) are separated. The eigen values of H^{tr} are the same as in case of the vacuum [cf. (16.45)]:

$$H_N = \sum_k h\, c\, |k| \cdot (N_k^{(2)} + N_k^{(3)} + 1).$$

If on the other hand the Hamilton operator $(H^P + H^C)$ of the "particle system," for instance, an atom or molecule with the stationary states $M = 1, 2, \ldots,$ has the eigen values E_M, then the unperturbed eigen function has the form:

$$F(t, N, M) = F(0, N, M) \cdot e^{-\frac{i t}{h}(H_N + E_M)}$$

The interaction term H^W describes now, in the well-known way, transitions between the states of the unperturbed system, i.e., simultaneous transitions of the field and of the material system. The probabilities of these transitions are determined by the respective matrix elements of H^W. According to (17.44) and (16.41) H^W (17.47) can be written as a matrix with respect to the quantum numbers $N_k^{(2)}, N_k^{(3)}$:

$$(17.49) \quad H^W = -\sum_k \sqrt{\frac{h\, c}{2\, V\, |k|}} \sum_{r=2,3} (a_k^{(r)} + a_{-k}^{(r)*}) \sum_n e_n\, (\alpha_n \cdot \mathrm{e}_k^{(r)})\, e^{i\, k\, x_n}.$$

Hence the matrix elements which determine the transitions are:

$$I^W_{N'\,M',\,N''\,M''} = -\sum_k \sqrt{\frac{h\, c}{2\, V\, |k|}} \sum_{r=2,3} (a_k^{(r)} + a_{-k}^{(r)*})_{N',\,N''} \cdot \left(\sum_n e_n\, (\alpha_n \cdot \mathrm{e}_k^{(r)})\, e^{i\, k\, x_n}\right)_{M',\,M''}$$

Since the operators $a_k^{(r)}$, $a_k^{(r)*}$ have matrix elements for transitions which decrease or increase the quantum numbers $N_k^{(r)}$ by 1, the perturbation function H^W describes light absorption and emission processes which are connected with the simultaneous changes of state of the material system. Hence perturbation theory yields in the first approximation the probabilities of the simple absorption

[1] Comprehensive articles on this field are: Fermi, *Rev. Modern Phys.* 4, 87, 1932; Breit, *ibid.* 4, 504, 1932, and 5, 91, 1933; Wentzel, *Handbuch der Physik*, Geiger-Scheel, Bd. 24, I. Teil, p. 740, 1933; Heitler, *Quantum Theory of Radiation*, Oxford Univ. Press, London, 1936; Kramers, *Hand- and Jahrbuch der Chemischen Physik*, Eucken-Wolf, Bd. I, Abschnitt II, Kap. 8, 1938.

and emission processes, in second approximation those of two-step processes, as, for instance, those of light scattering, etc.

As an example we shall discuss the scattering of a light quantum on a free electron, restricting ourselves to the unrelativistic limiting case, i.e., the velocity of the electron shall be neglected as small compared with the velocity of light. This shall be assumed not only for the initial state, but also for the final and intermediate states. This is only permitted if the Compton recoil is sufficiently small or in other words the momentum of the light quantum $h|k|$ must be chosen $\ll mc$. We have for the second order perturbation matrix:

$$H''_{\mathrm{FI}} = \sum_{\mathrm{II}} \frac{H^W_{\mathrm{F\,II}} \; H^W_{\mathrm{II\,I}}}{E_{\mathrm{I}} - E_{\mathrm{II}}},$$

where the indices I, II, F [as in (7.10)] refer to initial, intermediate and final states and where E_{I}, E_{II}, are the respective unperturbed energy eigen values. In the unrelativistic case one needs to consider only those elements of the matrix H^W, which correspond to transitions of the electron from the positive into the negative energy spectrum, or vice versa. All other matrix elements are, on account of the factor α, easily seen to be negligibly small. Thus, if the electron has in the initial and final states a positive energy ($\cong mc^2$), its energy in the intermediate state is negative[1] ($\cong - mc^2$). Since, on the other hand, we have assumed $hc|k| \ll mc^2$, the denominator $E_{\mathrm{I}} - E_{\mathrm{II}}$ in H''_{FI} has approximately the value $+2\,mc^2$ for all intermediate states which can be realized. In our approximation, we get:

$$H''_{\mathrm{F\,I}} = \frac{1}{2\,m\,c^2} \sum_{\mathrm{II}} H^W_{\mathrm{F\,II}} H^W_{\mathrm{II\,I}}$$

i.e., the matrix H'' is equal to the square of the matrix H^W divided by $2mc^2$. Instead of using the perturbation matrix H^W in second approximation, it is sufficient to discuss the perturbation function:

$$H'' = \frac{1}{2\,m\,c^2} \, (H^W)^2$$

in first approximation. In case of a single electron the operator H^W is according to (17.47):

$$H^W = -e\left(\alpha \cdot \psi^{\mathrm{tr}}(x_1)\right);$$

[1] If the electron waves are quantized according to the Pauli principle (theory of holes), the interpretation of the intermediate states will be somewhat different, cf. §21.

For its square one obtains on account of the well-known relations $\alpha^{(j)} \alpha^{(j')} + \alpha^{(j')} \alpha^{(j)} = 2 \delta_{jj'}$: $(H^W)^2 = e^2 [\psi^{tr} (x_1)]^2$, so that:

$$H'' = \frac{e^2}{2 m c^2} \left(\psi^{tr} (x_1)\right)^2.$$

It is interesting to note that the unrelativistic perturbation function (17.48) leads to the same result. In fact the term quadratic in ψ, which produces already a scattering in first approximation (Dirac's true scattering), agrees with H'', while the linear terms in ψ contribute only a negligible scattering in second approximation. With (17.44) and (16.41) we obtain:

$$H'' = \frac{h e^2}{4 m c V} \sum_{k k'} e^{i (k + k') x_1} \frac{1}{\sqrt{|k| |k'|}} \sum_{r, r' (\pm 1)} (\mathrm{e}_k^{(r)} \cdot \mathrm{e}_{k'}^{(r')}) \cdot (a_k^{(r)} + a_{-k}^{(r)*}) (a_{k'}^{(r')} + a_{-k'}^{(r')*}).$$

The matrix element for the transition of a light quantum from the state k, r into the state k', r' is thus equal to:

$$\frac{h e^2}{2 m c V} \frac{1}{\sqrt{|k| |k'|}} (\mathrm{e}_k^{(r)} \cdot \mathrm{e}_{k'}^{(r')}).$$

If one uses this to compute the intensity and the polarization of the scattered radiation, one will find complete agreement with the classical (Thomson) theory of light scattering.

§ 18. Multiple-Time Formalism

The fundamental equations of quantum electrodynamics which have been used so far, in particular the Schrödinger equation (17.17), do not contain the space and time coordinates in a symmetrical manner. For each electron we introduce a special space vector x_n and besides this the field coordinates x, while there enters only a single time coordinate t. In a Lorentz invariant formulation of the theory, one should assign to each particle its own individual "particle time" t_n, so that the space-time coordinates (x_n, t_n) of each single particle form a world vector, together with the field coordinates (x, t), which also form a world vector. The latter appear as arguments of the field functions. Dirac, Fock, and Podolsky[1] have found such an invariant formulation of

[1] *Physik. Z. Sowjetunion* 2, 468, 1932.

quantum electrodynamics which reduces to the theory with onetime coordinate—as represented in §17—if one lets all particle times t_n coincide with the field time t.

We shall use the time dependent operators $\psi_\nu(x, t)$, defined by (16.18). We denote the Hamilton operator of the vacuum field, which was called H in §16 [cf. (16.13)], now by H^0 as in equation (17.7):

$$\psi_\nu(x, t) = e^{\frac{it}{h} H^0} \psi_\nu(x) e^{-\frac{it}{h} H^0}. \tag{18.1}$$

The characteristic properties of these operators are the following: first, they satisfy as space-time functions the vacuum equations, i.e., the homogeneous wave equations:[1]

$$\Box \psi_\nu(x, t) = 0; \tag{18.2}$$

Furthermore, the invariant commutation rules (16.19) are:

$$[\psi_\nu(x, t), \psi_{\nu'}(x', t')] = \frac{h}{i} c^2 \delta_{\nu\nu'} D(x - x', t - t'). \tag{18.3}$$

With the help of these potential operators we define a Hamilton operator for each single electron:

$$H_n = m_n c^2 \beta_n + c\left(\alpha_n \cdot \left\{p_n - \frac{e_n}{c} \psi(x_n, t_n)\right\}\right) - e_n \cdot i \psi_4(x_n, t_n) \tag{18.4}$$

(ψ without index denotes again a 3-vector ψ_1, ψ_2, ψ_3).

Let Φ now be a Schrödinger function, which shall depend, not only on the field variables q and the electron variables q_n (= space vector x_n and spin coordinates; cf. §17), but also upon the individual particle times t_n.

$$\Phi = \Phi(q; t_1, q_1; t_2, q_2; \ldots); \tag{18.5}$$

on the other hand it shall be independent of the field time t, as well as of the field place x:

$$\frac{\partial \Phi}{\partial t} = 0, \qquad \nabla \Phi = 0. \tag{18.6}$$

[1] We draw attention to the difference between these operators and the operators (4.4) with $E = H$; for the last, according to (17.19), the inhomogeneous wave equations (17.5) are valid. With $E = H^0$, on the other hand, one can simply write, according to (16.16):

$$\frac{\partial^2 \psi_\nu(x, t)}{\partial t^2} = e^{\frac{it}{h} H^0} \cdot \left(\frac{i}{h}\right)^2 \left[H^0, [H^0, \psi_\nu(x)]\right] \cdot e^{-\frac{it}{h} H^0} = c^2 \nabla^2 \psi_\nu(x, t),$$

as mentioned in §16.

To determine how this function depends on the various particle times, one needs as many equations as there are particles. Thus, for each index n we postulate:

$$\left(\frac{h}{i}\frac{\partial}{\partial t_n} + H_n\right)\Phi = 0 \tag{18.7}$$

These simultaneous differential equations are not necessarily simultaneously integrable.[1] If one forms for one pair of values $n \neq n'$ with (18.7) the double time derivative:

$$\frac{\partial^2}{\partial t_{n'}\,\partial t_n}\Phi = \frac{\partial}{\partial t_{n'}}\left(-\frac{i}{h}H_n\Phi\right) = -\frac{i}{h}H_n\frac{\partial\Phi}{\partial t_{n'}} = \left(\frac{i}{h}\right)^2 H_n H_{n'}\Phi,$$

the result must be independent of the order of the two differentiations, the operators H_n and $H_{n'}$ must commute if the two corresponding equations (18.7) are to be compatible:

$$[H_n, H_{n'}] = 0. \tag{18.8}$$

Since the operators (18.4) contain the potentials ψ_ν obeying the commutation rules (18.3), it follows:

$$[H_n, H_{n'}] = \frac{h}{i}c^2 e_n e_{n'}\left\{(\alpha_n \cdot \alpha_{n'}) - 1\right\} D(x_n - x_{n'}, t_n - t_{n'});$$

These equations are only compatible in the space-time regions for which $D(x_n - x_{n'}, t_n - t_{n'})$ vanishes. According to (4.32) this equation is satisfied if the particles n and n' are situated spacelike to each other:

$$c\,|t_n - t_{n'}| < |x_n - x_{n'}|. \tag{18.9}$$

Within this limitation for all pairs $n \neq n'$, the differential equations (18.7) can be used to determine Φ. It may be noted that the single-time theory falls within the domain (18.9) of the integrability condition. The equations (18.7) are of course compatible with $\partial\Phi/\partial t = 0$ (18.6), since the operators H_n do not contain the field time t.

Since the differential equations (18.7) have the form of the Dirac wave equations [cf. (18.4)], they are evidently Lorentz-invariant. Under Lorentz transformations each of the equations (18.7) preserves its form if Φ is multiplied with an operator $\prod_n S_n$, where S_n is a matrix with respect to the spin coordinates

[1] Bloch, *Physik. Z. Sowjetunion* 5, 301, 1934.

of the nth electron, known from the one-electron problem. The region of validity of the Dirac equations is determined by the inequalities (18.9) also in an invariant way; each electron world point (x_n, t_n) must lie outside the light cones of all other electrons.

We claim now that the Schrödinger function F of the theory with the single time coordinate (§17) can be obtained from the function Φ, if one multiplies the latter with the operator $e^{-\frac{it}{h}H^0}$ and sets all particle times in it equal to the field time:

$$F(t, q, q_1, q_2, \ldots) = e^{-\frac{it}{h}H^0} \cdot \Phi(q; t, q_1; t, q_2; \ldots). \tag{18.10}$$

The time derivative of this function is namely:

$$\frac{\partial F}{\partial t} = e^{-\frac{it}{h}H^0} \cdot \left\{ -\frac{i}{h} H^0 \Phi + \sum_n \frac{\partial \Phi}{\partial t_n} \right\}_{t_1 = t_2 = \ldots = t},$$

according to (18.7):

$$-\frac{h}{i} \cdot \frac{\partial F}{\partial t} = e^{-\frac{it}{h}H^0} \cdot \left\{ \left(H^0 + \sum_n H_n \right) \Phi \right\}_{t_1 = t_2 = \ldots = t} \tag{18.11}$$

We move the factor $e^{-\frac{it}{h}H^0}$ towards the left of the operator $\left(H^0 + \sum_n H_n\right)$ and notice that it commutes with all terms except with the potential operators $\psi_\nu(x_n, t)$ appearing in $(H_n)_{t_n = t}$. According to (18.1) we have:

$$e^{-\frac{it}{h}H^0} \psi_\nu(x_n, t) = \psi_\nu(x_n)\, e^{-\frac{it}{h}H^0};$$

Thus the time-independent operators $\psi_\nu(x_n)$ replace the time-dependent operators. On the right side in (18.11), then, appears the operator, which was written in (17.9, 17.11) as $H_{(n)}$:

$$e^{-\frac{it}{h}H^0} (H_n)_{t_n = t} = H_{(n)}\, e^{-\frac{it}{h}H^0},$$

and one obtains finally with the help of (18.10):

$$-\frac{h}{i} \frac{\partial F}{\partial t} = \left(H^0 + \sum_n H_{(n)} \right) F \equiv H F. \tag{18.12}$$

This is now the Schrödinger equation (17.17), and our statement is proven.

We have thus shown that the simultaneous differential equations (18.7) represent a relativistic invariant generalization of the Schrödinger equation (17.17).

In order to obtain the Maxwell field equations in the theory with a single time coordinate, we subjected the function F in §17 to the subsidiary condition:

$$\pi_4(x)\cdot F(t, q, \ldots) = 0 \tag{18.13}$$

[cf. (17.21)]; the second subsidiary condition (17.21):

$$\dot{\pi}_4(x)\cdot F(t, q, \ldots) = 0$$

follows from the first; cf. (16.28) where $H = H^0 + \sum_n H_{(n)}$. If we express F according to (18.10) by Φ, (18.13) can be written after multiplication with $e^{\frac{it}{h}H^0}$:

$$e^{\frac{it}{h}H^0}\pi_4(x)\, e^{-\frac{it}{h}H^0}\cdot\Phi_{t=t_1=t_2=\ldots} = 0. \tag{18.14}$$

According to (17.18):

$$-ic\pi_4(x) = \nabla\cdot\psi(x) + \frac{1}{ic}\frac{i}{h}[H, \psi_4(x)],$$

or, since $H - H^0\left(=\sum_n H_{(n)}\right)$ commutes with $\psi_4(x)$:

$$-ic\pi_4(x) = \nabla\cdot\psi(x) + \frac{1}{ic}\frac{i}{h}[H^0, \psi_4(x)].$$

Substituting this into the equation (18.14), one can see by virtue of (18.1), that it is equivalent to:

$$\sum_\nu \frac{\partial\psi_\nu(x, t)}{\partial x_\nu}\cdot\Phi_{t=t_1=t_2=\ldots} = 0. \tag{18.15}$$

In a space with the coordinates $t, t_1, t_2, \ldots$ the "straight line" $t = t_1 = t_2 = \ldots$ represents the domain of validity of the single-time theory, i.e., with respect to changes of Φ along this line the simultaneous differential equations (18.7) are equivalent with the single-time Schrödinger equation for the function F (18.10). Since this Schrödinger equation according to §17 is compatible with the subsidiary condition (18.13), the equations (18.6, .7) and (18.15) must also be compatible along the single-time line. We shall now extend the equation (18.15) to a subsidiary condition for the multiple-time Schrödinger function Φ which shall be also valid outside this line. We assume:

$$\Omega\ \Phi = 0, \tag{18.16}$$

where Ω is an operator which reduces to $\sum_\nu \partial\psi_\nu(x, t)/\partial x_\nu$ for $t_n = t$. If one advances from any point of the single-time line along any curve into the domain of the multiple-time theory, the change of Φ is given by the equations (18.6, 18.7); Ω must be chosen in such a way, that $\Omega\,.\,\Phi$ along the path remains identically zero. Since each "point" $t, t_1, t_2, \ldots$ can be reached by advancing on the single-time line as far as the respective t-value and then by moving further in the "hyper-plane" $t =$ const., it is sufficient to consider path elements with $dt = 0$; i.e., it is sufficient to stipulate:

$$\frac{\partial}{\partial t_n}(\Omega\ \Phi) = 0 \text{ for all } n.$$

But according to (18.7):

$$\frac{\partial}{\partial t_n}(\Omega\ \Phi) = \left(\frac{\partial\Omega}{\partial t_n} - \frac{i}{h}\,\Omega\, H_n\right)\Phi \tag{18.17}$$

$$= \left(\frac{\partial\Omega}{\partial t_n} + \frac{i}{h}\,[H_n, \Omega] - \frac{i}{h}\,H_n\,\Omega\right)\Phi.$$

We shall now choose Ω in such a way that identically:

$$\frac{\partial\Omega}{\partial t_n} + \frac{i}{h}\,[H_n, \Omega] = 0 \quad (\text{for all } n) \tag{18.18}$$

Hence according to (18.17):

$$\frac{\partial}{\partial t_n}(\Omega\ \Phi) = -\frac{i}{h}\,H_n\,(\Omega\ \Phi);$$

i.e., if the function $\Omega\ \Phi$ vanishes at one point, it remains zero on all path elements which start from this point on the hyper-plane $t =$ const. Thus the equation (18.18) guarantees the validity of the subsidiary condition (18.16) in the whole domain, in which the differential equation (18.7) can be integrated, provided that Ω on the single-time line agrees with $\sum_\nu \partial\psi_\nu(x, t)/\partial x_\nu$ and that the single-time Schrödinger function (18.10) corresponds to the subsidiary conditions (17.21), i.e., to the initial conditions (17.20).

One can write instead of (18.18):

$$\left[\left\{\frac{\partial}{\partial t_n} + \frac{i}{h}\,H_n\right\}, \Omega\right] = 0; \tag{18.19}$$

where according to (18.4):

$$\frac{\partial}{\partial t_n} + \frac{i}{h} H_n = \frac{i}{h} m_n c^2 \beta_n + c\left(\alpha_n \cdot \left\{\nabla_n - \frac{i}{h}\frac{e_n}{c}\psi(x_n, t_n)\right\}\right) + \left\{\frac{\partial}{\partial t_n} - \frac{i}{h} e_n i \psi_4 (x_n, t_n)\right\}.$$

If we assume that Ω is independent of the spin coordinates of the electrons, or in other words that it commutes with the matrices α_n, β_n, then (18.19) is satisfied provided that Ω commutes with the operators:

$$\left\{\nabla_n - \frac{i}{h}\frac{e_n}{c}\psi(x_n, t_n)\right\} \quad \text{and} \quad \left\{\frac{1}{ic}\frac{\partial}{\partial t_n} - \frac{i}{h}\frac{e_n}{c}\psi_4(x_n, t_n)\right\}$$

i.e., if:

$$(18.20) \qquad \frac{\partial \Omega}{\partial x_{n\mu}} - \frac{e_n}{c}\frac{i}{h}[\psi_\mu(x_n, t_n), \Omega] = 0 \qquad (\mu = 1 \ldots 4,\ x_{n4} = i c t_n).$$

These conditions are satisfied with the following expression for Ω:

$$(18.21) \qquad \Omega = \sum_\nu \frac{\partial \psi_\nu (x, t)}{\partial x_\nu} + c \sum_n e_n D(x - x_n, t - t_n);$$

for, according to the commutation rules (18.3), we get:

$$\frac{i}{h}[\psi_\mu(x_n, t_n), \Omega] = -c^2 \frac{\partial}{\partial x_\mu} D(x - x_n, t - t_n)$$
$$= +c^2 \frac{\partial}{\partial x_{n\mu}} D(x - x_n, t - t_n),$$

so that (18.20) is satisfied identically. Since $D(x - x_n, 0)$ vanishes according to (4.31), we have also:

$$(\Omega)_{t = t_1 = t_2 = \ldots} = \sum_\nu \frac{\partial \psi_\nu (x, t)}{\partial x_\nu},$$

The desired multiple-time generalization of the subsidiary condition (18.13) is thus given by (18.16) in connection with (18.21). The operator (18.21) is evidently invariant under Lorentz transformations.

If we consider now the equations (18.2 to 18.7, and 18.16 with 18.21) as the fundamental equations of quantum electrodynamics—in fact all other statements can be derived from them—then the relativistic invariant character of the theory is evident. If one considers in any system of reference particularly the changes of the Schrödinger function Φ along the "line" $t = t_1 = t_2 \ldots$, one returns to the single-time theory treated in §17.

This formalism also permits gauge transformations:

$$\varphi_\nu (x, t) \rightarrow \varphi_\nu (x, t) + \frac{\partial \Lambda (x, t)}{\partial x_\nu},$$

where Λ is a space-time function (commuting with the ψ_ν), which satisfies the wave equation $\Box \Lambda = 0$, so that the equations (18.2, 18.3) are conserved.[1] If one transforms the multiple-time Schrödinger function according to:

$$\Phi \rightarrow e^{\frac{i}{hc} \sum_n e_n \Lambda (x_n, t_n)} \cdot \Phi,$$

the gauge invariance of the Dirac equations (18.7,.4) follows in the same way as the gauge invariance of the Dirac equation for one electron in a given field. The subsidiary condition (18.16, 18.21) is also gauge invariant (as consequence of $\Box \Lambda = 0$).

While the single-time Schrödinger function F of the ordinary wave mechanics has a known physical meaning—it determines the probability for obtaining certain experimental values of physical quantities—the multiple-time function Φ in the above formalism seems at first sight to be only a mathematical quantity. Bloch[2] showed, however, that Φ can also be interpreted as a probability amplitude for certain measurements, namely, those measurements (or series of measurements) which are made at the time t on the electromagnetic field, and which are taken at the times t_1, t_2, ... on the particles $n = 1, 2, \ldots$ The inequalities (18.9) which guarantee that the equations (18.7) can be integrated state that measurements do not disturb each other if they are carried out on different particles, since their actions on the field propagate at most with the velocity of light. Bloch's interpretation, furthermore, is only possible under the assumption that the field measurement is confined to such space-time regions (x, t), which are not influenced by the measurement of the particles and their reactions on the field. With these restrictive conditions $\Phi^*\Phi \, dq \, dq_1 \, dq_2 \ldots$ signifies the probability that one can find the field coordinates q in the domain dq at the time t as well as the coordinates q_1 of the particle $n = 1$ in the domain dq_1 at the time t_1, etc.

[1] Cf. footnote, page 113.

[2] Cf. footnote 1, p. 140.

If one considers any quantities:

$$f = f(x, t, q, q_1, q_2, \ldots),$$

one can define their "expectation values" as follows:

$$\int dq \int dq_1 \int dq_2 \ldots \Phi^* f \Phi = \bar{f}(x, t, t_1, t_2, \ldots).$$

According to (18.7) we have then:

$$\frac{\partial \bar{f}}{\partial t_n} = \frac{i}{h} \overline{[H_n, f]}, \tag{18.22}$$

while according to (18.6)

$$\frac{\partial \bar{f}}{\partial t} = \overline{\frac{\partial f}{\partial t}}, \qquad \nabla \bar{f} = \overline{\nabla f}. \tag{18.23}$$

With the help of (18.10) it is easy to see, that for $t = t_1 = t_2 = \ldots, \bar{f}$ reduces to the expectation value of the single-time theory, while f may stand for a particle quantity (such as, for instance, p_n) or a field quantity [for instance, $\nabla \times \psi(x, t)$]. One obtains, with (18.22), for the time changes of the single-time expectation values:

$$\begin{aligned} \frac{\partial}{\partial t} \bar{f}(x, t, t, t, \ldots) &= \left(\frac{\partial \bar{f}}{\partial t} + \sum_n \frac{\partial \bar{f}}{\partial t_n} \right)_{t_n = t} \\ &= \left(\overline{\frac{\partial f}{\partial t}} + \frac{i}{h} \sum_n \overline{[H_n, f]} \right)_{t_n = t}. \end{aligned} \tag{18.24}$$

If we apply these formulas, for instance, to the field strengths:

$$f_{\mu\nu} = \frac{\partial \psi_\nu(x, t)}{\partial x_\mu} - \frac{\partial \psi_\mu(x, t)}{\partial x_\nu}$$

then it follows from (18.2 and 18.21):

$$\begin{aligned} \sum_\mu \frac{\partial f_{\mu\nu}}{\partial x_\mu} &= -\frac{\partial}{\partial x_\nu} \sum_\mu \frac{\partial \psi_\mu(x, t)}{\partial x_\mu} \\ &= -\frac{\partial}{\partial x_\nu} \left\{ \Omega - c \sum_n e_n D(x - x_n, t - t_n) \right\}, \end{aligned}$$

and, since from (18.16) $\bar{\Omega} = 0$, we have:

$$\sum_{\mu} \frac{\partial \bar{f}_{\mu\nu}}{\partial x_\mu} = c \overline{\sum_n e_n \frac{\partial}{\partial x_\nu} D(x - x_n, t - t_n)}. \tag{18.25}$$

For the particular index $\nu = 4$, this equation states [cf. (16.1)]:

$$\nabla \cdot \bar{\mathfrak{E}} = \overline{\sum_n e_n \frac{\partial}{\partial t} D(x - x_n, t - t_n)}; \tag{18.26}$$

If all t_n approach the value t, one obtains, from (4.33):

$$\nabla \cdot \left(\bar{\mathfrak{E}}\right)_{t_n = t} = \sum_n e_n \left(\overline{\delta(x - x_n)}\right)_{t_n = t} = \left(\overline{\varrho(x)}\right)_{t_n = t},$$

in accordance with the single-time theory. On the other hand, (18.25) yields for $\nu = 1, 2, 3$:

$$\frac{1}{c}\frac{\partial \bar{\mathfrak{E}}}{\partial t} - \nabla \times \bar{\mathfrak{H}} = c \nabla \overline{\sum_n e_n D(x - x_n, t - t_n)}. \tag{18.27}$$

The right side vanishes for $t_n = t$. In order to obtain the corresponding Maxwell equations of the single-time theory, one must, according to (18.24), add the term:

$$\frac{1}{c}\frac{i}{h}\sum_n \overline{[H_n, \mathfrak{E}]}_{t_n = t}$$

For this term one finds after a short calculation, with the help of (18.3, 18.4), the following value:

$$-\sum_n e_n \left(\overline{\alpha_n \cdot \delta(x - x_n)}\right)_{t_n = t} = -\frac{1}{c}\left(\overline{s(x)}\right)_{t_n = t},$$

i.e.,

$$\frac{1}{c}\frac{\partial}{\partial t}(\bar{\mathfrak{E}})_{t_n = t} - \nabla \times (\bar{\mathfrak{H}})_{t_n = t} = -\frac{1}{c}\left(\overline{s(x)}\right)_{t_n = t},$$

which verifies, also, this Maxwell equation. The remaining Maxwell equations follow finally from the identities:

$$\nabla \cdot \mathfrak{H} = 0, \quad \frac{1}{c}\frac{\partial \mathfrak{H}}{\partial t} + \nabla \times \mathfrak{E} = 0 \tag{18.28}$$

if one takes into account that $[H_n, \mathfrak{H}]$ vanishes for $t_n = t$.

Of course, the above equations can also serve to follow the field $\mathfrak{E}$, $\mathfrak{H}$ into the domain outside the single-time domain. We shall carry this out for the field of one electron, but for reasons of simplicity only for the static limiting case; i.e., we assume that the electron is infinitely heavy and at rest, so that in H_1 (18.4) the matrices α and β may be replaced respectively by 0 and 1:

$$H_1 = \text{const.} - e \cdot i\, \psi_4 (x_1, t_1). \tag{18.29}$$

In the limit we may also assume that the position x_1 of the particle is sharply defined, for instance, at $x_1 = 0$; $\Phi^*\Phi$ then contains the factor $\delta(x_1)$, and constructing the expectation value with respect to x_1 simply means that one substitutes for x_1 the value 0. Thus, for instance:

$$\overline{D(x - x_1, t - t_1)} = D(x, t - t_1).$$

Hence, in the single-time case, Maxwell's equations hold, i.e., the laws of electrostatics:

$$\overline{\mathfrak{H}} = 0, \quad \overline{\mathfrak{E}} = -\frac{e}{4\pi} \nabla \frac{1}{|x|} \quad \text{for } t_1 = t. \tag{18.30}$$

Since $\mathfrak{H} = \nabla \times \psi(x, t)$ commutes with H_1 (18.29), it follows according to (18.22) that $\partial\mathfrak{H}/\partial t_1 = 0$; $\mathfrak{H}$ consequently vanishes not only for $t_1 = t$ but in general:

$$\overline{\mathfrak{H}} = 0. \tag{18.31}$$

According to (18.28) we have seen $\nabla \times \mathfrak{E} = 0$, hence $\mathfrak{E}$ can be represented as the gradient of a scalar function:

$$\overline{\mathfrak{E}} = -e \nabla U; \tag{18.32}$$

In this equation the potential function U may still depend on x and $(t - t_1)$. According to (18.30):

$$U = \frac{1}{4\pi |x|} \ (+ \text{const.}) \quad \text{for } t_1 = t. \tag{18.33}$$

If one substitutes (18.32) into the equations (18.26,.27), it follows:

$$\begin{cases} \nabla^2 U = -\dfrac{\partial}{\partial t} D(x, t - t_1), \\ \dfrac{\partial U}{\partial t} = -c^2 D(x, t - t_1); \end{cases} \tag{18.34}$$

The two equations are compatible with each other, since the Jordan-Pauli D-function satisfies the homogeneous wave equation: $\Box D = 0$. Applying the integral representation (16.20) for the D-function, one sees immediately that the two differential equations (18.34) are integrated by:

$$U = \frac{1}{(2\pi)^3} \int \frac{dk}{|k|^2} \cos\left[|k|\, c\, (t - t_1)\right] \cdot \cos k\, x. \tag{18.35}$$

For the evaluation of this integral we introduce polar coordinates in k-space: let ϑ be the angle which the vector k forms with the (constant) vector x; and $\cos\vartheta = \xi$; we write κ for $|k|$ so that the volume element becomes

$$dk = 2\pi\kappa^2\, d\kappa\ \ d\xi:$$

$$U = \frac{1}{(2\pi)^2} \int_0^\infty d\kappa \cos\left[\kappa\, c\, (t - t_1)\right] \int_{-1}^{+1} d\xi \cos\left(\kappa\, |x|\, \xi\right)$$

$$= \frac{1}{|x|} \frac{1}{(2\pi)^2} \int_0^\infty \frac{d\kappa}{\kappa}\, 2 \cos\left[\kappa\, c\, (t - t_1)\right] \sin\left(\kappa\, |x|\right)$$

$$= \frac{1}{|x|} \frac{1}{(2\pi)^2} \int_0^\infty \frac{d\kappa}{\kappa} \Big[\sin\left\{\kappa\left[|x| - c\, (t - t_1)\right]\right\}$$

$$+ \sin\left\{\kappa\left[|x| + c\, (t - t_1)\right]\right\}\Big].$$

Here the known integral appears:

$$I(\alpha) \equiv \int_0^\infty \frac{d\kappa}{\kappa} \sin\kappa\,\alpha = \begin{cases} +\dfrac{\pi}{2} & \text{for } \alpha > 0, \\[2ex] -\dfrac{\pi}{2} & \text{for } \alpha < 0; \end{cases}$$

and hence it follows:

$$U = \frac{1}{|x|} \cdot \frac{1}{(2\pi)^2} \left[I\left(|x| - c\, (t - t_1)\right) + I\left(|x| + c\, (t - t_1)\right)\right]. \tag{18.36}$$

If the 4-vector $x - 0$, $t - t_1$ is spacelike ($|x| > c|t - t_1|$), both arguments of the I-function will be positive; consequently we get:

$$U = \frac{1}{4\pi|x|} \quad \text{for } |x| > c\,|t - t_1|. \tag{18.37}$$

This agrees with (18.33) for $t_1 = t$. We see from (18.37) that U is independent of

$t - t_1$ everywhere outside the light cone $|x - x_1| = c|t - t_1|$ and is there equal to the Coulomb potential $1/4\pi\,|x - x_1|$. If we consider on the other hand the world points (x, t), which are situated timelike to the world point $(0, t_1)$ ($|x| < c|t - t_1|$, i.e., $t - t_1 > |x|/c$ or $t - t_1 < -|x|/c$), then the arguments of the I-functions in (18.36) and with it the values of the functions, too, have opposite signs, so that:

$$U = 0 \quad \text{for } |x| < c\,|t - t_1|; \tag{18.38}$$

Thus the field $\overline{\mathfrak{E}}, \overline{\mathfrak{H}}$ vanishes identically inside the light cone $|x - x_1| = c|t-t_1|$. Summarizing, we have for the static limiting case, from (18.31, 18.32, 18.37, 18.38):

$$\overline{\mathfrak{H}} = 0, \qquad \overline{\mathfrak{E}} = \begin{cases} -\dfrac{e}{4\pi}\,\nabla\,\dfrac{1}{|x|} & \text{for } |x| > c\,|t - t_1|, \\ 0 & \text{for } |x| < c\,|t - t_1|. \end{cases} \tag{18.39}$$

As an insertion we shall use here the formulas (18.37, 18.38) in order to discuss the Jordan-Pauli D-function as space-time function. According to (18.34) it is (with $t_1 = 0$):

$$D(x, t) = -\frac{1}{c^2}\frac{\partial U}{\partial t}, \quad \text{where } U = \begin{cases} \dfrac{1}{4\pi\,|x|} & \text{for } |x| > c\,|t|, \\ 0 & \text{for } |x| < c\,|t|. \end{cases} \tag{18.40}$$

Since U is independent of t outside as well as inside the light cone:

$$D(x, t) = 0 \quad \text{for } |x| \neq c\,|t|. \tag{18.41}$$

On the light cone itself, however, U is discontinuous, consequently $\partial U/\partial t$ becomes infinite, and this in such a way that $\int dt\, \partial U/\partial t$ (integrated over the singularity) $= \pm\, 1/4\pi\,|x|$; in other words, the integral becomes finite. Hence, the D-function has a δ-type singularity on the light cone. If we introduce a one-dimensional δ-function by:

$$\delta_1(t) = 0 \quad \text{for } t \neq 0, \qquad \int dt\, \delta_1(t) = 1,$$

we can write for the D-function:

$$D(x, t) = \frac{1}{4\pi c^2}\frac{1}{|x|}\left\{\delta_1\left(t - \frac{|x|}{c}\right) - \delta_1\left(t + \frac{|x|}{c}\right)\right\}. \tag{18.42}$$

for it agrees on the one hand with (18.41) for $|t| \neq |x|/c$; and on the other hand it leads with (18.40) to:

$$\frac{\partial U}{\partial t} = -c^2 D = \begin{cases} -\dfrac{1}{4\pi|x|}\,\delta_1\left(t-\dfrac{|x|}{c}\right) & \text{for } t \cong +\dfrac{|x|}{c}, \\ +\dfrac{1}{4\pi|x|}\,\delta_1\left(t+\dfrac{|x|}{c}\right) & \text{for } t \cong -\dfrac{|x|}{c}; \end{cases}$$

If one now passes, at constant distance $|x|$, the positive, or negative, light cone ($t = \pm|x|/c$) in the sense of increasing time (cf. the arrows in the

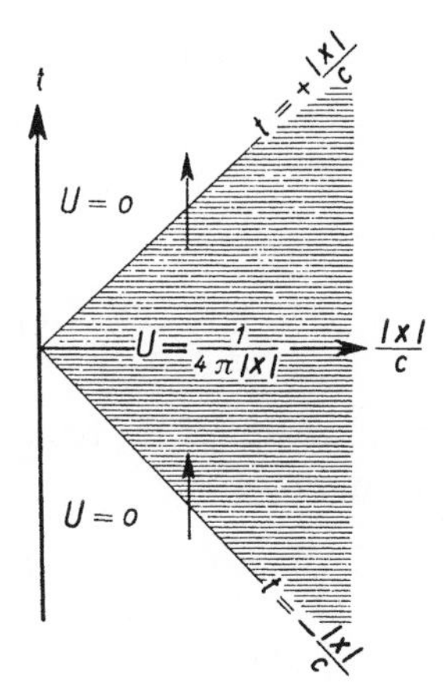

accompanying figure), we get $\int dt\, \partial U/\partial t = \pm 1/4\pi|x|$, in accordance with the discontinuities of U, which are given by (18.40). One should notice that the representation (18.42) is only valid for the Jordan-Pauli D-function, and not for the invariant functions (4.25) with $\mu \neq 0$; these vanish, according to (4.32), outside the light cone, but not inside it.

We state without proof the generalization of the formulas (18.39) regarding the field $\overline{\mathfrak{E}}$, $\overline{\mathfrak{H}}$ produced by an electron for the general, non-static case.[1] In the domain of the single-time theory all Maxwell equations may be integrated with the help of the retarded potentials:

$$\text{(18.43)} \qquad \overline{\mathfrak{H}} = \nabla\times\mathfrak{A}, \qquad \overline{\mathfrak{E}} = -\frac{1}{c}\frac{\partial\mathfrak{A}}{\partial t} - \nabla U,$$

$$\text{(18.44)} \qquad \begin{cases} U = U^{\text{ret}} \equiv \dfrac{1}{4\pi}\displaystyle\int dx'\, \frac{\bar{\varrho}\left(x', t - \dfrac{|x-x'|}{c}\right)}{|x-x'|}, \\ \mathfrak{A} = \mathfrak{A}^{\text{ret}} \equiv \dfrac{1}{4\pi c}\displaystyle\int dx'\, \frac{\bar{s}\left(x', t - \dfrac{|x-x'|}{c}\right)}{|x-x'|}. \end{cases}$$

[1] Wentzel, *Z. Phys.* *86*, 479, 1933.

It turns out that these field formulas are also valid for $t \neq t_1$[1] as long as the world point of the electron (x_1, t_1) is with certainty situated spacelike to the field point (x, t) ($\Phi \neq 0$ only in such x_1-domains for which $|x_1 - x| > c|t_1 - t|$, for given x, t and t_1). If one allows t to increase, *ceteris paribus*, until the field point (x, t) lies with certainty inside the (positive) light cone of the electron, the field $\overline{\mathfrak{E}}, \overline{\mathfrak{H}}$ vanishes as in the static case, i.e., we have in (18.43):

$$U = 0, \qquad \mathfrak{A} = 0 \text{ for } t - t_1 > \frac{|x - x_1|}{c}. \tag{18.45}$$

On the other hand, the field does not vanish if t tends towards $-\infty$; in this case we have (18.43) with:

$$U = U^{\text{ret}} - U^{\text{av}}, \quad \mathfrak{A} = \mathfrak{A}^{\text{ret}} - \mathfrak{A}^{\text{av}} \text{ for } t - t_1 < -\frac{|x - x_1|}{c}, \tag{18.46}$$

i.e., U and $\mathfrak{A}$ appear as the differences of the retarded and advanced potentials which are defined by (18.44) and by:

$$\left\{ \begin{aligned} U^{\text{av}} &\equiv \frac{1}{4\pi} \int dx' \frac{\bar{\varrho}\left(x', t + \frac{|x - x'|}{c}\right)}{|x - x'|}, \\ \mathfrak{A}^{\text{av}} &\equiv \frac{1}{4\pi c} \int dx' \frac{\bar{s}\left(x', t + \frac{|x - x'|}{c}\right)}{|x - x'|}. \end{aligned} \right. \tag{18.47}$$

The difference between retarded and advanced potentials disappears, of course, in the static limiting case, so that in this case, one falls back to (18.39).

§ 19. Remarks on the Self-Energy Problem

The multiple-time formalism has, on the one hand, the advantage that the Lorentz invariance of the theory is immediately obvious; on the other hand, it deserves a special interest in connection with the difficulty of the self-energy. Although it does not seem possible to remove this difficulty completely without a radical change of the theory (cf. §23), it is not without interest to study partial solutions of the problem. Here the multiple-time formalism opens up new possibilities.

For illustration we remind the reader of the corresponding problems in the classical (Lorentz) theory of the electron. For the limiting case of a point electron,

[1] In the multiple-time theory the $\bar{\rho}$ and $\bar{s}$ are functions of x and t_1, but they are independent of t [cf. 18.23) and (17.15)]; for t_1 one must substitute in the integrand of (18.44) the retarded time $t - |x - x'|/c$.

the field energy and with it the mass of the electron becomes infinite. At the same time the self-force, i.e., the force which an electron experiences on account of the action of its own field, becomes infinite, too. This self-force is obtained from the limiting values of the field strengths of the electron's own field at the position of the electron and these limiting values are undetermined in the multiple-time formalism. It must be added that the transition to the classical limiting case ($h \to 0$) is easily carried out in both the single-time as well as the multiple-time formalisms. In particular, the formulas (18.43 to 18.47) for the field produced by an electron remain valid, where only the density functions $\bar{\rho}$, $\bar{s}$ in the potentials (18.44, 18.47) are replaced by the classical density functions (17.3). The retarded potentials (18.44) are then identical with the well-known Liénard-Wiechert potentials of a moving point charge. The same holds for the advanced potentials (18.47). In order to determine the limiting value of the field strengths at the position of the electron (x_1, t_1) one lets the "field point" (x, t) approach (x_1, t_1). In the single-time theory one has beforehand $t_1 = t$ and it is well known that in this case the limiting value for the field strengths is not finite. The same holds true if one approaches the particle position from an arbitrary spacelike direction

$$\left(\frac{c|t - t_1|}{|x - x_1|} < 1\right).$$

Inside the positive light cone ($t > t_1$), on the other hand, the field strengths vanish according to (18.45) and consequently also their limiting values at the position of the electron. Inside the negative light cone ($t < t_1$) the field strengths are not zero and they approach a finite limiting value since the singularities in the retarded and advanced potentials just cancel each other in (18.46). This allows the possibility of a classical theory of a point electron and its field in which the resulting self-forces are finite.[1] In order to obtain conservation of energy it is necessary to define the field strength at the position as the arithmetical mean of the limiting values which are obtained for the approach from the positive and negative light cone. The self-force reduces then to the Lorentz force ($e^2/6\pi c^3 \cdot \dddot{x}_1$ in the rest system of the particle). This means there exists, in contradistinction to Lorentz's extended model, no electromagnetic inertia force. The self-energy also can be made finite by suitable definition.

The above mentioned rule for evaluating the limiting value at the position of the electron can also be taken over into quantum theory.[2] According to Dirac[3] it is also possible to modify the commutation rules (18.3) by replacing the invariant D-function by a modified non-invariant function which later is made to approach the invariant D-function. We shall adopt this formulation here.

Let ξ represent a small vector in x-space and τ a small time interval such that ξ and $c\tau$ form a timelike 4-vector:

$$c^2\tau^2 - \xi^2 = s^2 > 0. \tag{19.1}$$

Later we shall let ξ and τ tend to zero in such a way that the direction of the vector $\xi/c\tau$ remains a constant. We denote this limiting process, in short, as the "limiting

---

[1] Wentzel, *Z. Phys. 86*, 479, 1933; *87*, 726, 1934. Another equivalent formulation was given by Dirac, *Proc. Roy. Soc. London 167*, 148, 1938.

[2] Wentzel, *Z. Phys. 86*, 635, 1933.

[3] Dirac, *Ann. Inst. Henri Poincaré 9*, 13, 1939. (Lecture given in Paris, March 1939)

process $s \to 0$".[1] We use now the Jordan-Pauli D-function (16.20) in order to define a new space-time function which depends on the parameters ξ, τ:

$$D_s(x, t) = \frac{1}{2}\left\{D(x + \xi, t + \tau) + D(x - \xi, t - \tau)\right\}. \tag{19.2}$$

This function is not Lorentz-invariant since it distinguishes the reference system in which the space components ξ of the 4-vector vanish. In the limiting case, $s \to 0$, it is equal to the invariant D-function.

We adopt from §18 the fundamental equations of the multiple-time theory, but replace everywhere D by D_s, following Dirac's example [cf. (18.2 to 18.7, 18.16, 18.21)]:

$$\Box \psi_\nu(x, t) = 0, \tag{19.3}$$

$$[\psi_\nu(x, t), \psi_{\nu'}(x', t')] = \frac{h}{i} c^2 \delta_{\nu\nu'} D_s(x - x', t - t'), \tag{19.4}$$

$$\left[\left\{\frac{h}{i}\frac{\partial}{\partial t_n} - e_n i \psi_4(x_n, t_n)\right\} + c\left(\alpha_n \cdot \left\{\frac{h}{i}\nabla_n - \frac{e_n}{c}\psi(x_n, t_n)\right\}\right) + m_n c^2 \beta_n\right]\Phi = 0, \tag{19.5}$$

$$\left\{\sum_\nu \frac{\partial \psi_\nu(x, t)}{\partial x_\nu} + c\sum_n e_n D_s(x - x_n, t - t_n)\right\}\Phi = 0. \tag{19.6}$$

These "subsidiary conditions" (19.6) are compatible with the Dirac equation (19.5), which may be seen by a calculation similar to the one in §18 [cf. (18.19) ff.]. The proof carried out there did not make use of the special dependence of the D-function on space and time but was based solely on the fact that the same D-function appears in the subsidiary conditions as in the commutation rules, which is also true here. With the new commutation rules (19.4) one finds for the integrability condition of the Dirac equations (19.5) [cf. (18.8)]:

$$c|t_n - t_{n'} + \tau| < |x_n - x_{n'} + \xi|, \qquad c|t_n - t_{n'} - \tau| < |x_n - x_{n'} - \xi| \tag{19.7}$$

(for all pairs $n \neq n'$). The domain of integrability is thus slightly smaller than (18.9). In the limit $s \to 0$, however, it regains its old extension.

In order to show how the new commutation rules (19.4) can be satisfied together with (19.3), we shall give an explicit matrix representation of the operators $\psi_\nu(x, t)$, which satisfies these conditions. The wave equations (19.3) hold if we write the ψ_ν as superpositions of plane waves:

$$\psi_\nu(x, t) = V^{-1/2}\sum_k \left\{A_{\nu,k}\, e^{i(kx - |k|ct)} + A^*_{\nu,k}\, e^{-i(kx - |k|ct)}\right\}; \tag{19.8}$$

Let:

$$A_{\nu,k} = \left[\frac{hc}{2|k|}\cos(k\xi - |k|c\tau)\right]^{1/2} \cdot a_{\nu,k}, \quad A^*_{\nu,k} = \left[\frac{hc}{2|k|}\cos(k\xi - |k|c\tau)\right]^{1/2} \cdot a^*_{\nu,k}, \tag{19.9}$$

[1] Dirac (cf. footnote 3, p. 153) uses for the 4-vector s the letter λ and the limiting process here discussed is in the literature often referred to as the λ-limiting process.

where $a_{\nu,k}$, $a^*_{\nu,k}$ are matrices of the kind repeatedly used with the commutation rules:

$$\left[a_{\nu,k}, a_{\nu',k'}\right] = \left[a^*_{\nu,k}, a^*_{\nu',k'}\right] = 0, \quad \left[a_{\nu,k}, a^*_{\nu',k'}\right] = \delta_{\nu\nu'}\,\delta_{kk'}.$$ [1]

The equations (19.8, 19.9) then yield:

$$[\psi_\nu(x,t), \psi_{\nu'}(x',t')] = i\,h\,c\,\delta_{\nu\nu'}\cdot V^{-1}\sum_k \frac{1}{|k|}\cos(k\,\xi - |k|\,c\,\tau)\cdot\sin\left\{k\,(x-x') - |k|\,c\,(t-t')\right\}.$$

If one replaces here (corresponding to the limiting case $V \rightarrow \infty$) $V^{-1}\sum_k \ldots$ by $(2\pi)^{-3}\int dk \ldots$ then an elementary calculation confirms the equation (19.4) [in connection with (19.2) and (16.20)].

[1] It should be noted here that $a^*_{4,k}$ (on account of the imaginary character of ψ_4) denotes the operator which is the Hermitian conjugate to $-a_{4,k}$. If we write:

$$a_{4,k} = i\,a_{0k}, \quad a^*_{4,k} = i\,a^*_{0k},$$

the commutation rules for the matrices a_{0k} and a^*_{0k}, which are Hermitian conjugates, are:

$$\left[a_{0k}, a^*_{0k}\right] = -1 \quad \text{or} \quad \left[a^*_{0k}, a_{0k}\right] = +1;$$

The roles of the matrices a and a^* are thus interchanged compared with the components $\nu = 1, 2, 3$.

In the limiting case $s \rightarrow 0$, $\cos(k\xi - |k|c\tau) \rightarrow 1$, the equation (19.8, 19.9) corresponds to the definition (18.1) of the time-dependent potential operators, where:

$$\psi_\nu(x) = \psi_\nu(x, 0) = V^{-1/2}\sum_k \sqrt{\frac{h\,c}{2\,|k|}}\left(a_{\nu,k} + a^*_{\nu,-k}\right)e^{ikx}.$$

To verify this, one may use that according to (19.18), (16.11), and (17.7):

$$H^0 = h\,c\sum_k |k| \sum_\nu a^*_{\nu,k}\,a_{\nu,k} + \text{const}$$

consequently:

$$H^0\,a_{\nu,k} = a_{\nu,k}\,(H^0 - h\,c\,|k|), \qquad H^0\,a^*_{\nu,k} = a^*_{\nu,k}\,(H^0 + h\,c\,|k|),$$

$$e^{\frac{it}{h}H^0}\,a_{\nu,k} = a_{\nu,k}\,e^{\frac{it}{h}(H^0 - h\,c\,|k|)}, \qquad e^{\frac{it}{h}H^0}\,a^*_{\nu,k} = a^*_{\nu,k}\,e^{\frac{it}{h}(H^0 + h\,c\,|k|)}$$

and finally:

$$e^{\frac{it}{h}H^0}\,\psi_\nu(x) = \psi_\nu(x,t)\,e^{\frac{it}{h}H^0},$$

in accordance with (18.1).

In order to evaluate the subsidiary condition (19.6), we decompose the 3-vector a_k (components $a_{1,k}$, $a_{2,k}$, $a_{3,k}$) and a_k^* into longitudinal and transverse components:

$$a_k = \sum_r \mathfrak{e}_k^{(r)} a_k^{(r)}, \qquad a_k^* = \sum_r \mathfrak{e}_k^{(r)} a_k^{(r)*}; \tag{19.10}$$

here the coordinate axes $\mathfrak{e}_k^{(r)}$ are again defined by the formulas (12.29, 12.30), ($\mathfrak{e}_k^{(1)} = +k/|k|$) and the commutation rules of the $a_k^{(r)}$, $a_k^{(r)*}$ are given by (16.43). Thus we get:

$$\sum_\nu \frac{\partial \varphi_\nu (x, t)}{\partial x_\nu} = i\, V^{-1/2} \sum_k \left[\frac{h\, c\, |k|}{2} \cos (k\, \xi - |k|\, c\, \tau)\right]^{1/2} \cdot \left\{\left(a_k^{(1)} + i\, a_{4,k}\right) e^{i\,(k\,x - |k|\,c\,t)} - \left(a_k^{(1)*} + i\, a_{4,k}^*\right) e^{-i\,(k\,x - |k|\,c\,t)}\right\}.$$

On the other hand, it follows with (19.2) and (16.20):

$$c \sum_n e_n\, D_s (x - x_n, t - t_n) = \frac{i}{2\,V} \sum_k \frac{1}{|k|} \cos (k\, \xi - |k|\, c\, \tau) \cdot \left\{e^{i\,(k\,x - |k|\,c\,t)} \sum_n e_n\, e^{-i\,(k\,x_n - |k|\,c\,t_n)} - e^{-i\,(k\,x - |k|\,c\,t)} \sum_n e_n\, e^{i\,(k\,x_n - |k|\,c\,t_n)}\right\}.$$

In order to satisfy the subsidiary conditions (19.6) identically in x and t, it is evidently necessary that:

$$\left(a_k^{(1)} + i\, a_{4,k} + C_k\right) \Phi = 0, \qquad \left(a_k^{(1)*} + i\, a_{4,k}^* + C_k^*\right) \Phi = 0, \tag{19.11}$$

where:

$$C_k = \left[\frac{1}{2\, h\, c\, V} \cos (k\, \xi - |k|\, c\, \tau)\right]^{1/2} |k|^{-3/2} \sum_n e_n\, e^{-i\,(k\,x_n - |k|\,c\,t_n)}. \tag{19.12}$$

If we write:

$$\left\{\begin{aligned} P_k &= h\left(a_k^{(1)} + i\, a_{4,k}\right), & P_k^* &= h\left(a_k^{(1)*} + i\, a_{4,k}^*\right), \\ Q_k &= \frac{1}{2\,i}\left(a_k^{(1)*} - i\, a_{4,k}^*\right), & Q_k^* &= -\frac{1}{2\,i}\left(a_k^{(1)} - i\, a_{4,k}\right), \end{aligned}\right. \tag{19.13}$$

then the Q_k, Q_k^* and P_k, P_k^* are canonically conjugate, since:

$$\left[P_k, Q_k\right] = \left[P_k^*, Q_k^*\right] = \frac{h}{i},$$

while the other pairs commute. Writing the Schrödinger function Φ as a function of the Q_k and Q_k^*, then $P_k = -ih\partial/\partial Q_k$, $P_k^* = -ih\partial/\partial Q_k^*$, and we obtain for (19.11) with (19.13):

$$\left(\frac{1}{i} \frac{\partial}{\partial Q_k} + C_k\right) \Phi = 0, \qquad \left(\frac{1}{i} \frac{\partial}{\partial Q_k^*} + C_k^*\right) \Phi = 0.$$

These differential equations may be integrated by:

$$\Phi = S \cdot \Phi^{\mathrm{tr}}, \tag{19.14}$$

where:

$$S = e^{-i \sum_k (C_k Q_k + C_k^* Q_k^*)} \tag{19.15}$$

and where Φ^{tr} is independent of the Q_k and Q_k^*:

$$\frac{\partial \Phi^{\mathrm{tr}}}{\partial Q_k} = \frac{\partial \Phi^{\mathrm{tr}}}{\partial Q_k^*} = 0. \tag{19.16}$$

Introducing (19.14) into the Dirac equations (19.5), we must again move the operator S towards the left (cf. §17). First of all it follows with (19.15):

$$\frac{h}{i} \frac{\partial S}{\partial t_n} = -h \sum_k \left(\frac{\partial C_k^*}{\partial t_n} Q_k^* + \frac{\partial C_k}{\partial t_n} Q_k \right) \cdot S,$$

or with (19.12):

$$\frac{h}{i} \frac{\partial S}{\partial t_n} = i\, e_n \left(\frac{h c}{2 V} \right)^{1/2} \sum_k \left[\frac{1}{|k|} \cos (k \xi - |k|\, c \tau) \right]^{1/2} \cdot$$
$$\cdot \left\{ Q_k^* \, e^{i (k x_n - |k|\, c t_n)} - Q_k \, e^{-i (k x_n - |k|\, c t_n)} \right\} \cdot S.$$

Furthermore, according to (19.8, 19.9 and 19.13):

$$-e_n\, i\, \psi_4 (x_n, t_n) = -i\, e_n \left(\frac{h c}{2 V} \right)^{1/2} \sum_k \left[\frac{1}{|k|} \cos (k \xi - |k|\, c \tau) \right]^{1/2} \cdot$$
$$\cdot \left\{ \left(Q_k^* - \frac{i}{2h} P_k \right) e^{i (k x_n - |k|\, c t_n)} - \left(Q_k + \frac{i}{2h} P_k^* \right) e^{-i (k x_n - |k|\, c t_n)} \right\},$$

hence:

$$\frac{h}{i} \frac{\partial S}{\partial t_n} - e_n\, i\, \psi_4 (x_n, t_n)\, S = i\, e_n \left(\frac{h c}{2 V} \right)^{1/2} \sum_k \left[\frac{1}{|k|} \cos (k \xi - |k|\, c \tau) \right]^{1/2} \cdot$$
$$\cdot \frac{1}{2} \left\{ e^{i (k x_n - |k|\, c t_n)} \frac{\partial}{\partial Q_k} + e^{-i (k x_n - |k|\, c t_n)} \frac{\partial}{\partial Q_k^*} \right\} S.$$

Considering (19.14, 19.15, 19.16) we have:

$$\left\{ \frac{h}{i} \frac{\partial}{\partial t_n} - e_n\, i\, \psi_4 (x_n, t_n) \right\} \Phi = S \left\{ \frac{h}{i} \frac{\partial}{\partial t_n} + e_n\, Y_n \right\} \Phi^{\mathrm{tr}}, \tag{19.17}$$

where:

$$Y_n = \left(\frac{h c}{2 V} \right)^{1/2} \sum_k \left[\frac{1}{|k|} \cos (k \xi - |k|\, c \tau) \right]^{1/2} \cdot \frac{1}{2} \left\{ e^{i (k x_n - |k|\, c t_n)} C_k \right.$$
$$\left. + e^{-i (k x_n - |k|\, c t_n)} C_k^* \right\}, \tag{19.18}$$

or with (19.12):

$$Y_n = \frac{1}{2V}\sum_k \frac{1}{|k|^2}\cos(k\,\xi - |k|\,c\,\tau)\cdot\sum_{n'} e_{n'}\cos\{k\,(x_{n'} - x_n) - |k|\,c\,(t_{n'} - t_n)\}.$$

With the help of the function:

$$(19.19)\quad U(x,t) = \frac{1}{V}\sum_k \frac{1}{|k|^2}\cos(k\,x - |k|\,c\,t) = \frac{1}{(2\pi)^3}\int dk\,\frac{1}{|k|^2}\cos(k\,x - |k|\,c\,t)$$

Y_n can evidently be represented as follows:

$$(19.20)\quad Y_n = \frac{1}{4}\sum_{n'} e_{n'}\{U(x_{n'} - x_n + \xi,\ t_{n'} - t_n + \tau) + U(x_{n'} - x_n - \xi,\ t_{n'} - t_n - \tau)\}.$$

$U(x,t)$ is identical with the function (18.35), if one puts in the latter $t_1 = 0$. According to (18.37, 18.38) it follows:

$$(19.21)\quad U(x,t) = \begin{cases} \dfrac{1}{4\pi|x|} & \text{for } c\,|t| < |x|, \\ 0 & \text{for } c\,|t| > |x|. \end{cases}$$

According to this the term of the sum with $n' = n$ in (19.20) vanishes, since the 4-vector ξ, $c\tau$ is supposed to be a timelike vector [cf. (19.1)]. For $n' \neq n$, on the other hand, the 4-vectors $x_{n'} - x_n \pm \xi$, $c(t_{n'} - t_n \pm \tau)$ are spacelike according to (19.7). Hence it follows:

$$(19.22)\quad Y_n = \frac{1}{4}\sum_{\substack{n' \\ (n' \neq n)}} \frac{e_{n'}}{4\pi}\left\{\frac{1}{|x_{n'} - x_n + \xi|} + \frac{1}{|x_{n'} - x_n - \xi|}\right\},$$

and in the limit $s \to 0$:

$$(19.23)\quad \lim_{s=0} Y_n = \frac{1}{2}\sum_{\substack{n' \\ (n' \neq n)}} \frac{e_{n'}}{4\pi\,|x_{n'} - x_n|};$$

This term represents half of the Coulomb potential of the other charges ($n' \neq n$) at the position of the nth electron. It is noteworthy that the infinite self-potential of the point charges does not appear due to the introduction of the timelike vector ξ, $c\tau$.

If, on the other hand, one decomposes the 3-potential $\psi(x,t)$ into longitudinal and transverse components:

$$(19.24)\quad \psi = \psi^{\text{long}} + \psi^{\text{tr}},$$

where ψ^{long} contains the terms $\sim a_k^{(1)}$ and $a_k^{(1)*}$, it follows with (19.13):

$$-\frac{e_n}{c}\,\psi^{\text{long}}(x_n, t_n) = i\,e_n\left(\frac{h}{2cV}\right)^{1/2}\sum_k \frac{k}{|k|}\left[\frac{1}{|k|}\cos(k\,\xi - |k|\,c\,\tau)\right]^{1/2}\cdot$$

$$\cdot\left\{\left(Q_k^* + \frac{i}{2h}P_k\right)e^{i(k\,x_n - |k|\,c\,t_n)} - \left(Q_k - \frac{i}{2h}P_k^*\right)e^{-i(k\,x_n - |k|\,c\,t_n)}\right\}.$$

Here the terms with Q_k^* and Q_k cancel again, if we form:

$$\frac{h}{i} \nabla_n S - \frac{e_n}{c} \psi^{\text{long}} (x_n, t_n) S,$$

and one obtains, analogous to (19.17), with (19.24):

$$\left\{\frac{h}{i} \nabla_n - \frac{e_n}{c} \psi (x_n, t_n)\right\} \Phi \tag{19.25}$$

$$= S\left\{\frac{h}{i} \nabla_n - \frac{e_n}{c} \psi^{\text{tr}} (x_n, t_n) + \frac{e_n}{c} \mathfrak{Z}_n\right\} \Phi^{\text{tr}},$$

$\mathfrak{Z}_n$ differs from Y_n (19.18) only by a factor $k/|k|$ under the summation sign. One has therefore the formulas analogous to (19.20 and .19):

$$\mathfrak{Z}_n = \frac{1}{4} \sum_{n'} e_{n'} \left\{\mathfrak{V} (x_{n'} - x_n + \xi, t_{n'} - t_n + \tau) + \mathfrak{V} (x_{n'} - x_n - \xi, t_{n'} - t_n - \tau)\right\}, \tag{19.26}$$

where:

$$\mathfrak{V} (x, t) = \frac{1}{V} \sum_{k} \frac{k}{|k|^3} \cos (k x - |k| c t). \tag{19.27}$$

By comparison with (19.19), one finds:

$$\mathfrak{V} (x, t) = -c \nabla \int_0^t dt\, U (x, t).$$

Since according to (19.21):

$$c \int_0^t dt\, U (x, t) = \begin{cases} \dfrac{1}{4\pi} \dfrac{c t}{|x|} & \text{for } c|t| < |x|, \\ \dfrac{1}{4\pi} & \text{for } c t > |x|, \\ -\dfrac{1}{4\pi} & \text{for } c t < -|x|, \end{cases}$$

it follows thus:

$$\mathfrak{V} (x, t) = \begin{cases} \dfrac{1}{4\pi} \dfrac{c t \cdot x}{|x|^3} & \text{for } c|t| < |x|, \\ 0 & \text{for } c|t| > |x|. \end{cases} \tag{19.28}$$

In $\mathfrak{Z}_n$ (19.26), according to this, the term of the sum $n' = n$ (on account of $c|\tau| > |\xi|$) vanishes again, the remaining terms being:

$$\mathfrak{Z}_n = \frac{c}{4} \sum_{\substack{n' \\ (n' \neq n)}} \frac{e_{n'}}{4\pi} \left\{\frac{(t_{n'} - t_n + \tau)(x_{n'} - x_n + \xi)}{|x_{n'} - x_n + \xi|^3} + \frac{(t_{n'} - t_n - \tau)(x_{n'} - x_n - \xi)}{|x_{n'} - x_n - \xi|^3}\right\}, \tag{19.29}$$

and in the limit $s \to 0$:

$$\lim_{s=0} \mathfrak{Z}_n = \frac{c}{2} \sum_{\substack{n' \\ (n' \neq n)}} \frac{e_{n'}}{4\pi} \frac{(t_{n'} - t_n)(x_{n'} - x_n)}{|x_{n'} - x_n|^3}. \tag{19.30}$$

According to (19.5, 19.17, 19.25) the Dirac equations for Φ^{tr} can be written as follows:

$$\left[\left\{\frac{h}{i}\frac{\partial}{\partial t_n} + e_n Y_n\right\} + c\left(\alpha_n \cdot \left\{\frac{h}{i}\nabla_n + \frac{e_n}{c}\mathfrak{Z}_n - \frac{e_n}{c}\psi^{\mathrm{tr}}(x_n, t_n)\right\}\right) + m_n c^2 \beta_n\right]\Phi^{\mathrm{tr}} = 0. \tag{19.31}$$

In these equations, besides the electron coordinates, only the transverse field components enter.

We shall, furthermore, investigate how Φ^{tr} changes, on account of (19.31), along the "single-time" line $t_1 = t_2 = \ldots = t$. If we write for short:

$$\Phi^{\mathrm{tr}}_{t_1 = t_2 = \ldots = t} = F'(t), \tag{19.32}$$

it follows, on account of:

$$\frac{\partial F'}{\partial t} = \left(\sum_n \frac{\partial \Phi^{\mathrm{tr}}}{\partial t_n}\right)_{t_1 = t_2 = \ldots = t},$$

with the notation:

$$H^P = \sum_n \left\{c\left(\alpha_n \cdot \frac{h}{i}\nabla_n\right) + m_n c^2 \beta_n\right\}$$

[cf. (17.12)] and:

$$H^C = \sum_n e_n \{Y_n + (\alpha_n \cdot \mathfrak{Z}_n)\}_{t_1 = t_2 = \ldots = t}: \tag{19.33}$$

$$\left\{\frac{h}{i}\frac{\partial}{\partial t} + H^P + H^C - \sum_n e_n\left(\alpha_n \cdot \psi^{\mathrm{tr}}(x_n, t)\right)\right\} F' = 0. \tag{19.34}$$

In the limiting case $s \to 0$ ($\tau \to 0$) the contribution of $\mathfrak{Z}_n$ to H^C vanishes according to (19.30). Thus according to (19.23):

$$\lim_{s=0} H^C = \frac{1}{8\pi} \sum_{n \neq n'} \frac{e_n e_{n'}}{|x_n - x_{n'}|}; \tag{19.35}$$

This is exactly the Coulomb energy of the electrons, without the self-potential terms in contradistinction to (17.39). We write, analogous to (18.10):

$$F^{\mathrm{tr}} = e^{-\frac{it}{h}H^{\mathrm{tr}}} F', \qquad F' = e^{\frac{it}{h}H^{\mathrm{tr}}} F^{\mathrm{tr}}$$

where H^{tr} represents the contribution of the transverse waves to the field energy H^0 *in vacuo* [cf. (17.34)]. Again one can move in (19.34) the factor $e^{\frac{it}{h}H^{tr}}$ to the left side by virtue of the formula:

$$\psi^{tr}(x_n, t)\, e^{\frac{it}{h}H^{tr}} = e^{\frac{it}{h}H^{tr}}\, \psi^{tr}(x_n)$$

Since $H^0 - H^{tr}$ commutes with ψ^{tr} we have according to (18.1):

$$\psi^{tr}(x, t) = e^{\frac{it}{h}H^0}\, \psi^{tr}(x)\, e^{-\frac{it}{h}H^0} = e^{\frac{it}{h}H^{tr}}\, \psi^{tr}(x)\, e^{-\frac{it}{h}H^{tr}}\Big).$$

One thus obtains for F^{tr} the differential equations (17.46, 17.47), where, however, H^C and the commutation rules for ψ^{tr} are changed according to (19.33) and (19.4).

Even without the self-potential terms, a self-energy of the electrons may still appear on account of their interaction with the transverse light waves. We shall investigate this problem, restricting the discussion to the case of a single electron. In this case we have only one Dirac equation (19.31) for Φ^{tr}, and Y_n and $\mathfrak{Z}_n$ vanish [there are no terms $n' \neq n$ in (19.22, 19.29)]. We introduce the 4-vectors:

$$(x_1,\ i c t_1) = \vec{x}, \qquad (i \alpha \beta, \beta) = \vec{\gamma},$$

and write for (19.31):

$$\left\{\left(\vec{\gamma} \cdot \frac{\vec{\partial}}{\partial x}\right) + \mu - \chi\right\} \Phi^{tr} = 0, \tag{19.36}$$

where:

$$\mu = \frac{m c}{h}, \tag{19.37}$$

$$\chi = \frac{i e}{h c}\left(\gamma \cdot \psi^{tr}\left(\vec{x}\right)\right). \tag{19.38}$$

We consider χ as perturbation function, i.e., we expand Φ^{tr} in powers of the parameter e:

$$\Phi^{tr} = \Phi^0 + \Phi' + \Phi'' + \ldots, \tag{19.39}$$

$$\begin{cases} \left\{\left(\vec{\gamma} \cdot \frac{\vec{\partial}}{\partial x}\right) + \mu\right\} \Phi^0 = 0, \\ \left\{\left(\vec{\gamma} \cdot \frac{\vec{\partial}}{\partial x}\right) + \mu\right\} \Phi' = \chi\, \Phi^0, \\ \left\{\left(\vec{\gamma} \cdot \frac{\vec{\partial}}{\partial x}\right) + \mu\right\} \Phi'' = \chi\, \Phi', \\ \ldots\ldots\ldots \end{cases} \tag{19.40}$$

Operating with $\left(\vec{\gamma} \cdot \frac{\vec{\partial}}{\partial x}\right) - \mu$ on these equations, one obtains, with the help of:

$$\gamma_\nu \gamma_{\nu'} + \gamma_{\nu'} \gamma_\nu = 2\,\delta_{\nu\nu'} \tag{19.41}$$

the following equations:

$$\begin{cases} (\square - \mu^2)\,\Phi^0 = 0, \\ (\square - \mu^2)\,\Phi' = \chi'\,\Phi^0, \\ (\square - \mu^2)\,\Phi'' = \chi'\,\Phi', \\ \dots\dots\dots, \end{cases} \tag{19.42}$$

where:

$$\chi' = \left\{\left(\vec{\gamma} \cdot \frac{\vec{\partial}}{\partial x}\right) - \mu\right\} \chi. \tag{19.43}$$

For the computation of the self-energy of the electron the unperturbed Schrödinger function Φ^0 may be chosen as a state of the system with no light quanta present:

$$\Phi^0_N = \begin{cases} e^{i(\vec{p}\cdot\vec{x})}u, \text{ if all } N_k^{(r)} = 0, \\ 0 \text{ otherwise.} \end{cases} \tag{19.44}$$

Here u is a spinor amplitude, which according to (19.40) satisfies the equation:

$$\{i\,(\vec{\gamma} \cdot \vec{p}) + \mu\}\, u = 0 \tag{19.45}$$

According to (19.42) one finds, of course:

$$\vec{p}^{\,2} = -\mu^2 \tag{19.46}$$

$\left(h\vec{p}\right.$ = energy-momentum vector of the unperturbed electron$\left.\right)$.

In order to compute Φ' according to (19.40 or 19.42), we must form $\chi\Phi^0$ and $\chi'\Phi^0$. By operating with ψ^{tr} [cf. (19.8, 19.9, 19.10)] on Φ^0 (19.44), we find that the operators $a_k^{(r)}$ give zero (for they correspond to absorption processes which cannot take place if no light quanta are present). Hence only the terms $\sim a_k^{(r)*}$ remain:

$$\psi^{tr}(\vec{x})\,\Phi^0 = \left(\frac{hc}{2V}\right)^{1/2} \sum_k \left[\frac{1}{|k|}\cos\left(\vec{k}\cdot\vec{\xi}\right)\right]^{1/2} \sum_{r=2,3} e_k^{(r)}\, a_k^{(r)*}\, e^{-i(\vec{k}\cdot\vec{x})}\,\Phi^0$$

$\left(\vec{k} = (k, i|k|),\ \vec{\xi} = (\xi, ic\tau)\right)$. In the expression $\chi'\Phi^0$, there occurs, according to (19.38 and 19.43), the differential operator $\frac{\vec{\partial}}{\partial x}$ which operates on the exponential function $e^{i(\vec{p}-\vec{k})\cdot\vec{x}}$, so that:

$$\chi'\,\Phi^0 = i\,e\,(2\,h\,c\,V)^{-1/2} \sum_k \left[\frac{1}{|k|}\cos\left(\vec{k}\cdot\vec{\xi}\right)\right]^{1/2} \left\{i\,\vec{\gamma}\cdot\left(\vec{p}-\vec{k}\right)-\mu\right\}\cdot$$

$$\sum_{r=2,3} \left(\gamma\cdot e_k^{(r)}\right) a_k^{(r)*}\, e^{-i\left(\vec{k}\cdot\vec{x}\right)}\,\Phi^0.$$

Since, on the other hand:

$$\begin{aligned}(\Box-\mu^2)\,e^{i\left(\vec{p}-\vec{k}\right)\cdot\vec{x}} &= -\left\{\left(\vec{p}-\vec{k}\right)^2+\mu^2\right\} e^{i\left(\vec{p}-\vec{k}\right)\cdot\vec{x}}\\ &= 2\left(\vec{p}\cdot\vec{k}\right) e^{i\left(\vec{p}-\vec{k}\right)\cdot\vec{x}}\end{aligned}$$

(on account of $\vec{p}^{\,2} = -\mu^2$ and $\vec{k}^{\,2} = 0$), it follows for Φ' according to (19.42):

$$\Phi' = \frac{i}{2}\,e\,(2\,h\,c\,V)^{-1/2} \sum_k \left[\frac{1}{|k|}\cos\left(\vec{k}\cdot\vec{\xi}\right)\right]^{1/2} \frac{1}{\left(\vec{p}\cdot\vec{k}\right)} \left\{i\,\vec{\gamma}\cdot\left(\vec{p}-\vec{k}\right)-\mu\right\}\cdot$$

$$\sum_{r=2,3} \left(\gamma\cdot e_k^{(r)}\right) a_k^{(r)*}\, e^{-i\left(\vec{k}\cdot\vec{x}\right)}\,\Phi^0.$$

Here we notice that by (19.41):

$$\left\{i\,\vec{\gamma}\cdot\left(\vec{p}-\vec{k}\right)-\mu\right\}\left(\gamma\cdot e_k^{(r)}\right) = -\left(\gamma\cdot e_k^{(r)}\right)\left\{i\,\vec{\gamma}\cdot\left(\vec{p}-\vec{k}\right)+\mu\right\} + 2\,i\,(p-k)\cdot e_k^{(r)},$$

or, since $k\cdot e_k^{(r)} = 0$ (for $r = 2, 3$), and with regard to (19.45):

$$\left\{i\,\vec{\gamma}\cdot\left(\vec{p}-\vec{k}\right)-\mu\right\}\left(\gamma\cdot e_k^{(r)}\right) u = i\left\{\left(\gamma\cdot e_k^{(r)}\right)\left(\vec{\gamma}\cdot\vec{k}\right) + 2\left(p\cdot e_k^{(r)}\right)\right\} u.$$

Therefore:

$$\Phi' = -\frac{1}{2}\,e\,(2\,h\,c\,V)^{-1/2} \sum_k \left[\frac{1}{|k|}\cos\left(\vec{k}\cdot\vec{\xi}\right)\right]^{1/2} \frac{1}{\left(\vec{p}\cdot\vec{k}\right)}\cdot \tag{19.47}$$

$$\sum_{r=2,3} \left\{\left(\gamma\cdot e_k^{(r)}\right)\left(\vec{\gamma}\cdot\vec{k}\right) + 2\left(p\cdot e_k^{(r)}\right)\right\} a_k^{(r)*}\, e^{-i\left(\vec{k}\cdot\vec{x}\right)}\,\Phi^0.$$

In order to calculate Φ'' in the same way, we must operate with $\psi^{tr}(x)$ once more on Φ'. Thus, on the one hand, there appear terms $\sim a_{k'}^{(r')*}\, a_k^{(r)*}\,\Phi^0$ which correspond to the presence of two light quanta (double emission), and, on the other hand, terms $\sim a_{k'}^{(r')}\, a_k^{(r)*}\,\Phi^0$, which are only different from zero for $k' = k$, $r' = r$:

$$a_k^{(r)}\, a_k^{(r)*}\,\Phi^0 = \Phi^0.$$

Only this latter term (Φ_0'') of Φ'' is of interest in this connection since it alone determines the self-energy terms $\sim e^2$. According to (19.8, 19.9, 19.10, 19.42, 19.43, and

19.47) this expression satisfies the equation:

$$(\Box - \mu^2)\, \Phi_0'' = -\frac{i}{4}\frac{e^2}{hc}\left\{\left(\vec{\gamma}\cdot\frac{\vec{\partial}}{\partial x}\right) - \mu\right\}\frac{1}{V}\sum_k \frac{1}{|k|}\cos\left(\vec{k}\cdot\vec{\xi}\right)\frac{1}{\left(\vec{p}\cdot\vec{k}\right)}\cdot$$

$$\sum_{r=2,3}\left(\gamma\cdot e_k^{(r)}\right)\left\{\left(\gamma\cdot e_k^{(r)}\right)\left(\vec{\gamma}\cdot\vec{k}\right) + 2\left(p\cdot e_k^{(r)}\right)\right\}\Phi^0,$$

or:

$$(\Box - \mu^2)\, \Phi_0'' = \lambda\cdot\Phi^0, \tag{19.48}$$

where:

$$\lambda = -\frac{i}{2}\frac{e^2}{hc}\left\{i\left(\vec{\gamma}\cdot\vec{p}\right) - \mu\right\}\frac{1}{V}\sum_k \frac{1}{|k|}\cos\left(\vec{k}\cdot\vec{\xi}\right)\frac{1}{\left(\vec{p}\cdot\vec{k}\right)}\cdot$$

$$\cdot\left\{\left(\vec{\gamma}\cdot\vec{k}\right) + \sum_{r=2,3}\left(\gamma\cdot e_k^{(r)}\right)\left(p\cdot e_k^{(r)}\right)\right\}[1] \tag{19.49}$$

(on account of $(\gamma\cdot e_k^{(r)})^2 = 1$). We claim that the terms $\sim \sum_r (\gamma\cdot e_k^{(r)})\,(p\cdot e_k^{(r)})$ are zero after the summation over k is carried out and can therefore be omitted. If one integrates first with respect to $|k| = \kappa$ for constant direction of the 3-vector k (i.e., with $e_k^{(r)}$ = const.), $(\vec{k}\cdot\vec{\xi})$ and $(\vec{p}\cdot\vec{k})$ vary proportionally to κ, and the integral $\int_0^\infty d\kappa \cos(\beta\kappa)$ appears. By introducing the factor $e^{-\alpha\kappa}$, we obtain for this integral:

$$\int_0^\infty d\kappa \cos(\beta\kappa)\, e^{-\alpha\kappa} = \frac{\alpha}{\alpha^2+\beta^2},$$

which vanishes in the limit $\alpha \to 0$. This result is based on the fact that the limiting process $\alpha \to 0$ is carried out before the limiting process $s \to 0$ (i.e., $\beta \to 0$). It is also essential that β should be different from zero for all directions of k. This is true on account of the timelike character of the 4-vector $\vec{\xi}$. The special κ-dependency of the factor $e^{-\alpha\kappa}$ which produces convergence is, however, not essential. The formula (19.49) reduces now to:

$$\lambda = -\frac{i}{2}\frac{e^2}{hc}\left\{i\left(\vec{\gamma}\cdot\vec{p}\right) - \mu\right\}\frac{1}{V}\sum_k \frac{1}{|k|}\cos\left(\vec{k}\cdot\vec{\xi}\right)\frac{\left(\vec{\gamma}\cdot\vec{k}\right)}{\left(\vec{p}\cdot\vec{k}\right)}.$$

If one considers further that from (19.41):

$$\left\{i\left(\vec{\gamma}\cdot\vec{p}\right) - \mu\right\}\left(\vec{\gamma}\cdot\vec{k}\right) = -\left(\vec{\gamma}\cdot\vec{k}\right)\left\{i\left(\vec{\gamma}\cdot\vec{p}\right) + \mu\right\} + 2i\left(\vec{p}\cdot\vec{k}\right)$$

[1] The single-time theory (§17) leads to the same formulas with $\xi = 0$. In this case the sum with respect to k in (19.49) diverges and (19.48) leads to the infinite self-energy of the electron, calculated first by Waller (*Z. Phys.* *62*, 673, 1930).

and that the operator $\left\{i\left(\vec{\gamma}\cdot\vec{p}\right)+\mu\right\}$ applied to Φ^0 yields zero according to (19.45), one finds that λ in (19.48) may be replaced by:

$$(19.50)\qquad \lambda = \frac{e^2}{h\,c}\,\frac{1}{V}\sum_k \frac{1}{|k|}\cos\left(\vec{k}\cdot\vec{\xi}\right) = \frac{e^2}{h\,c}\,\frac{1}{(2\,\pi)^3}\int\frac{dk}{|k|}\cos\left(k\,\xi - |k|\,c\,\tau\right).$$

The k-space integral can evidently be represented by the two Lorentz-invariant space-time functions (4.27)[1] (with $\mu = 0$, $x = \xi$, $t = \tau$). The quantity λ is thus invariant in the sense that it will depend only on the magnitude of the 4-vector $\vec{\xi}$, i.e., only on s [cf. (19.1)]. In order to evaluate λ, we may thus choose that system of reference in which the space components of $\vec{\xi}$ vanish.

$$\xi = 0,\quad c\,|\tau| = s,\quad \cos\left(\vec{k}\cdot\vec{\xi}\right) = \cos\left(|k|\,s\right) = \cos\left(\varkappa\,s\right).$$

Introducing again the factor $e^{-\alpha\varkappa}$, we find:

$$\int\frac{dk}{|k|}\cos\left(\vec{k}\cdot\vec{\xi}\right) = \lim_{\alpha=0} 4\pi\int_0^\infty d\varkappa\,\varkappa\cos\left(\varkappa\,s\right)e^{-\alpha\varkappa}$$

$$= \lim_{\alpha=0} 4\pi\,\frac{\alpha^2 - s^2}{(\alpha^2+s^2)^2} = -\frac{4\pi}{s^2},$$

consequently:

$$(19.51)\qquad \lambda = -\frac{1}{2\,\pi^2}\,\frac{e^2}{h\,c}\,\frac{1}{s^2}.$$

λ becomes infinite in the limit $s \to 0$ and from (19.48) the same holds true for Φ_0''. In its present form the theory is thus not yet free of infinities; but the infinity in the term $\sim e^2$ which was calculated here can be eliminated, as Dirac has pointed out, by a simple modification of the Hamiltonian of the electron. If we replace, in the Dirac equation (19.36), μ by $\mu + \eta$ where η is a small constant involving e^2, we obtain in place of the third equation (19.40 and 19.42):

$$\left\{\left(\vec{\gamma}\cdot\frac{\vec{\partial}}{\partial x}\right)+\mu\right\}\Phi'' = \chi\,\Phi' - \eta\,\Phi^0,$$

$$(\Box - \mu^2)\,\Phi'' = \chi'\,\Phi' - \eta\left\{\left(\vec{\gamma}\cdot\frac{\vec{\partial}}{\partial x}\right)-\mu\right\}\Phi^0$$

$$= \chi'\,\Phi' + 2\,\eta\,\mu\,\Phi^0,$$

and hence in place of (19.48):

$$(\Box - \mu^2)\,\Phi_0'' = (\lambda + 2\,\eta\,\mu)\,\Phi^0.$$

---

[1] Namely by the sum of the two functions, while the D-function (4.25) is determined by their difference.

If we choose:

$$\eta = -\frac{\lambda}{2\mu} = \frac{1}{4\pi^2}\frac{e^2}{hc}\frac{1}{\mu}\frac{1}{s^2}, \tag{19.52}$$

then Φ_0'' vanishes and the theory in the second perturbation approximation will no longer contain any infinities. As may be easily seen this is also true if several electrons are present and all rest masses m_n in the Dirac equations (19.5) are changed correspondingly. According to (19.37) m_n must be replaced by:

$$m_n(s) = m_n + \frac{1}{4\pi^2}\frac{e^2 h}{c^3}\frac{1}{m_n}\frac{1}{s^2}. \tag{19.53}$$

If one thus assigns to the field-free particles values for the rest masses $m_n(s)$, which in the limit $s \to 0$ become positive infinite, they will move in an electromagnetic field like electrons of the rest masses m_n. This may be interpreted in the sense that the "electromagnetic rest mass" $-\frac{1}{4\pi^2}\frac{e^2 h}{c^3}\frac{1}{m_n}\frac{1}{s^2}$ has been added to the "mechanical rest mass" $m_n(s)$.

This is a quantum theory of field and point charges which has no infinities in the terms $\sim e^2$ even in the limit $s \to 0$. Although the introduction of the timelike 4-vector ξ, $c\tau$ implies a distinction of the coordinate system in which the space components ξ vanish, the theory is, nevertheless, in the discussed approximation, in the limit $s \to 0$ invariant under Lorentz transformations.[1]

[1] It must be stressed, however, that this theory fails to eliminate the logarithmic divergencies which occur in Dirac's theory of the positron ("theory of holes," see §21). This failure has led Dirac to propose a new method of field quantization (*Proc. Roy. Soc. London 180*, 1, 1942), but this line of attack does not seem very promising, mainly in view of the difficulties in the physical interpretation. Cf. also Pauli, *Rev. Modern Phys. 15*, 175, 1943.

Chapter V

The Quantization of the Electron Wave Field According to the Exclusion Principle

§ 20. Force-Free Electrons

The description of the interaction between light and electrons which was discussed in the previous chapter does not put forth any analogies in the representation of light quanta and electrons. The quantum features of light appear as a consequence of the quantization of the light waves, the electrons on the other hand were introduced from the beginning as individual units and were treated according to the quantum mechanical method of the configuration space, by considering the Schrödinger function as a function of the coordinates of the different electrons. The situation is similar for the case of the mesons, on the one hand, and protons and neutrons, on the other hand, in the theory of the meson field (Chaps. II, III). A uniform description of particles with integral and non-integral spin can be achieved by subjecting also the wave fields of the electrons, protons, etc., to a process of quantization.[1] The exclusion principle of Pauli, which holds for these particles, necessitates certain modifications of the canonical formalism used so far. The "quantization according to the Bose-Einstein statistics" must be replaced by a "quantization according to the Pauli principle," or according to the "Fermi-Dirac statistics." To illustrate this further, we shall first discuss the force-free motion of the electrons.

[1] One speaks of a "second quantization" or "hyperquantization" because the transition from classical mechanics to wave mechanics corresponds to a first quantization. The analogy: light quantum-electron could, of course, also be expressed by describing the light quanta in the configuration space (cf. Dirac, *Proc. Roy. Soc. London 114*, 243, 1927; Pauli, *Handbuch der Physik*, Geiger-Scheel, Bd. *24*, I. Teil, p. 259); but the theory loses then much of its formal simplicity, especially if absorption and emission processes occur, in which the number of particles changes.

We start with the Dirac wave equation of the force-free electrons:

$$(20.1)\qquad \left(\frac{h}{i}\frac{\partial}{\partial t}+\boldsymbol{E}\right)\psi=0,\qquad \boldsymbol{E}=\frac{c\,h}{i}(\alpha\cdot\nabla)+m\,c^2\beta;$$

ψ signifies the 4-component spinor field; the components of the vector α and $\beta=\alpha^{(4)}$ are the Dirac matrices with the properties:

$$(20.2)\qquad \alpha^{(\nu)}=\alpha^{(\nu)*},\qquad \alpha^{(\mu)}\alpha^{(\nu)}+\alpha^{(\nu)}\alpha^{(\mu)}=2\,\delta_{\mu\nu}\,.$$

We consider the complex field components ψ_σ first as classical field functions. The Lagrangian of the problem can be written as follows:[1]

$$(20.3)\qquad \boldsymbol{L}=-\psi^*\left(\frac{h}{i}\frac{\partial}{\partial t}+\boldsymbol{E}\right)\psi=-\psi^*\left\{\frac{h}{i}\dot\psi+\frac{c\,h}{i}(\alpha\cdot\nabla\psi)+m\,c^2\beta\,\psi\right\}$$

$$=-\sum_\varrho\psi_\varrho^*\left\{\frac{h}{i}\dot\psi_\varrho+\frac{c\,h}{i}\sum_k\sum_\sigma\alpha_{\varrho\sigma}^{(k)}\frac{\partial\psi_\sigma}{\partial x_k}+m\,c^2\sum_\sigma\beta_{\varrho\sigma}\psi_\sigma\right\};$$

If $\boldsymbol{L}$ is considered as a function of the ψ_σ, ψ_σ^* and their derivatives, the variations of the ψ_σ yield according to (1.2):

$$\frac{h}{i}\dot\psi_\sigma^*+\frac{c\,h}{i}\sum_k\sum_\varrho\frac{\partial\psi_\varrho^*}{\partial x_k}\alpha_{\varrho\sigma}^{(k)}-m\,c^2\sum_\varrho\psi_\varrho^*\beta_{\varrho\sigma}=0,$$

which agrees with the complex conjugate of the equation (20.1). This equation is obtained directly by the variation of the ψ_σ^*, the derivatives of which do not appear in $\boldsymbol{L}$:

$$-\frac{\partial\boldsymbol{L}}{\partial\psi_\varrho^*}=\frac{h}{i}\dot\psi_\varrho+\frac{c\,h}{i}\sum_k\sum_\sigma\alpha_{\varrho\sigma}^{(k)}\frac{\partial\psi_\sigma}{\partial x_k}+m\,c^2\sum_\sigma\beta_{\varrho\sigma}\psi_\sigma=0.$$

Charge and current density of the field are according to (3.11) defined in the following way:

$$\varrho=-i\,\varepsilon\sum_\sigma\left(\frac{\partial\boldsymbol{L}}{\partial\dot\psi_\sigma}\psi_\sigma-\frac{\partial\boldsymbol{L}}{\partial\dot\psi_\sigma^*}\psi_\sigma^*\right)=\varepsilon\,h\sum_\sigma\psi_\sigma^*\psi_\sigma,$$

$$s_k=-i\,\varepsilon\sum_\sigma\left(\frac{\partial\boldsymbol{L}}{\partial\frac{\partial\psi_\sigma}{\partial x_k}}\psi_\sigma-\frac{\partial\boldsymbol{L}}{\partial\frac{\partial\psi_\sigma^*}{\partial x_k}}\psi_\sigma^*\right)=\varepsilon\,h\,c\sum_{\varrho\sigma}\psi_\varrho^*\alpha_{\varrho\sigma}^{(k)}\psi_\sigma,$$

[1] For reasons of simplicity we have chosen an unsymmetrical representation in ψ and ψ^*. This could easily be removed by the addition of a space-time divergence (cf. the footnote on page 2).

hence:

$$\varrho = \varepsilon\, h \cdot \psi^* \psi, \qquad s = \varepsilon\, h\, c \cdot \psi^* \alpha\, \psi, \tag{20.4}$$

in agreement with the known equations of the Dirac wave mechanics. With the help of the matrices:

$$\gamma^{(k)} = -i\, \beta\, \alpha^{(k)} = i\, \alpha^{(k)}\, \beta \quad (k = 1, 2, 3), \quad \gamma^{(4)} = \beta, \tag{20.5}$$

which also satisfy the relations:

$$\gamma^{(\nu)} = \gamma^{(\nu)*}, \qquad \gamma^{(\mu)}\gamma^{(\nu)} + \gamma^{(\nu)}\gamma^{(\mu)} = 2\,\delta_{\mu\nu}$$

and with the notation:

$$\psi^\dagger = i\, \psi^* \beta \tag{20.6}$$

the formulas (20.1, 20.3, 20.4) can be written:

$$\left(\sum_\nu \gamma^{(\nu)} \frac{\partial}{\partial x_\nu} + \mu\right)\psi = 0 \qquad \left(\mu = \frac{m\,c}{h}\right), \tag{20.7}$$

$$L = -\frac{h\,c}{i}\psi^\dagger\left(\sum_\nu \gamma^{(\nu)} \frac{\partial}{\partial x_\nu} + \mu\right)\psi, \tag{20.8}$$

$$s_\nu = \varepsilon\, h\, c \cdot \psi^\dagger \gamma^{(\nu)} \psi \qquad \left(\sum_\nu \frac{\partial s_\nu}{\partial x_\nu} = 0\right). \tag{20.9}$$

The canonical energy-momentum tensor constructed according to the rules (2.8) or (3.8) is:

$$T^0_{\mu\nu} = \frac{h\,c}{i}\psi^\dagger \gamma^{(\mu)} \frac{\partial \psi}{\partial x_\nu} + L\,\delta_{\mu\nu}.$$

Here the term $L\delta_{\mu\nu}$ vanishes on account of the validity of the Dirac equation (20.7). By adding the tensor:

$$T'_{\mu\nu} = -\frac{h\,c}{2\,i}\frac{\partial}{\partial x_\nu}\psi^\dagger \gamma^{(\mu)} \psi,$$

the divergence of which $\sum_\mu \partial T'_{\mu\nu}/\partial x_\mu$ vanishes according to the continuity equation for the 4-current, we obtain a tensor with the correct reality properties:[1]

$$T''_{\mu\nu} = \frac{hc}{2i}\left(\psi^\dagger \gamma^{(\mu)} \frac{\partial \psi}{\partial x_\nu} - \frac{\partial \psi^\dagger}{\partial x_\nu} \gamma^{(\mu)} \psi\right).$$

This tensor is not yet symmetrical; but since not only $\sum_\mu \partial T''_{\mu\nu}/\partial x_\mu = 0$, but—as can easily be verified—also $\sum_\nu \partial T''_{\mu\nu}/\partial x_\nu = 0$, the symmetrical tensor $T_{\mu\nu} = \frac{1}{2}(T''_{\mu\nu} + T''_{\nu\mu})$ also satisfies the conservation equation $\sum_\mu \partial T_{\mu\nu}/\partial x_\mu = 0$:

$$(20.10) \quad T_{\mu\nu} = \frac{hc}{4i}\left\{\psi^\dagger \gamma^{(\mu)} \frac{\partial \psi}{\partial x_\nu} + \psi^\dagger \gamma^{(\nu)} \frac{\partial \psi}{\partial x_\mu} - \frac{\partial \psi^\dagger}{\partial x_\nu} \gamma^{(\mu)} \psi - \frac{\partial \psi^\dagger}{\partial x_\mu} \gamma^{(\nu)} \psi\right\}.$$

Under Lorentz transformations $\bar{x}_\mu = \sum_\nu a_{\mu\nu} x_\nu$ the ψ and $\psi^\dagger$ are transformed according to $\bar{\psi} = S\psi$, $\bar{\psi}^\dagger = \psi^\dagger S^{-1}$, where the matrix S is determined by:

$$S^{-1} \gamma^{(\mu)} S = \sum_\nu a_{\mu\nu} \gamma^{(\nu)} \cdot \quad (\text{und } S^{-1} = \beta S^* \beta)$$

so that:

$$(20.11) \qquad \sum_\nu \gamma^{(\nu)} \frac{\partial}{\partial x_\nu} = \sum_{\nu\mu} \gamma^{(\nu)} a_{\mu\nu} \frac{\partial}{\partial \bar{x}_\mu} = S^{-1}\left(\sum_\mu \gamma^{(\mu)} \frac{\partial}{\partial \bar{x}_\mu}\right) S.$$

From this there follows directly the Lorentz invariance of the Lagrangian L (20.8) and also the vector character of s_ν (20.9) and the tensor character of $T_{\mu\nu}$ (20.10).

Since according to (20.3) the canonical momenta $\pi^*_\sigma = \partial L/\partial \dot{\psi}^*_\sigma$ vanish identically, we shall try to eliminate the field functions ψ^*_σ, π^*_σ before carrying out the transition to the Hamiltonian formalism, as in the case of the meson field (§12). This can be accomplished immediately with the help of the relations:

$$(20.12) \qquad \pi_\sigma = \frac{\partial L}{\partial \dot{\psi}_\sigma} = -\frac{h}{i} \psi^*_\sigma.$$

[1] That is, with real 4,4- and j,j'-components and with imaginary 4,j- and j,4-components. If one transforms the Lagrangian according to:

$$L \to L + \frac{hc}{2i} \sum_\nu \frac{\partial}{\partial x_\nu} (\psi^\dagger \gamma^{(\nu)} \psi)$$

(cf. footnote 1, p. 2), one obtains, as canonical energy-momentum tensor, the tenor T''. Cf. also Appendix I.

We obtain thus for the Hamiltonian $H = \sum_\sigma (\pi_\sigma \dot{\psi}_\sigma + \pi^*_\sigma \dot{\psi}^*_\sigma) - L \, (= - T^0_{44})$:

$$(20.13) \qquad H = -\frac{i}{h} \pi E \psi = -\sum_{\varrho\sigma} \pi_\varrho \left\{ c\,(\alpha_{\varrho\sigma} \cdot \nabla) + \frac{i m c^2}{h} \beta_{\varrho\sigma} \right\} \psi_\sigma .$$

It is easy to see, on account of (20.1, 20.2, and 20.12), that the integral Hamiltonian function $H = \int dx \, H$ is real, i.e., a Hermitian operator and that it is equal to the total energy $-\int dx \, T_{44}$ [cf. (20.10)]. If we wanted to carry out the quantization on the basis of the canonical commutation relations (1.7), we would obtain, from (20.13), the canonical field equations:

$$\dot{\psi} = \frac{i}{h} [H, \psi] = -\frac{i}{h} E \psi, \quad \dot{\pi} = -\frac{h}{i} \dot{\psi}^* = -(E\psi)^*,$$

formally in agreement with (20.1).

However, a first objection to this canonical quantization method is that then the total energy of the electrons is not positive-definite. In order to show this, we expand the ψ-function with respect to the eigen functions of the Dirac-equation (20.1), where we impose again spatial periodicity in order to obtain a discrete energy spectrum:[1]

$$(20.14) \quad \psi_\sigma(x, t) = \sum_m a_m e^{-\frac{it}{h} E_m} u_{m\sigma}(x), \quad \psi^*_\sigma(x, t) = \sum_m a^*_m e^{\frac{it}{h} E_m} u^*_{m\sigma}(x),$$

where:

$$(20.15) \qquad (-E_m + E)\, u_m = 0, \quad \int_V dx \, u^*_m u_{m'} = \delta_{mm'} .$$

According to (20.12 and 20.14) it follows that:

$$(20.16) \qquad \pi_\sigma(x, t) = -\frac{h}{i} \sum_m a^*_m e^{\frac{it}{h} E_m} u^*_{m\sigma}(x).$$

The commutation rules (1.7) are satisfied, if the expansion coefficients a_m, a^*_m are considered as matrices of the type (6.16) with the commutators:

$$[a_m, a_{m'}] = [a^*_m, a^*_{m'}] = 0, \quad [a_m, a^*_{m'}] = \delta_{mm'};$$

[1] The index m, enumerating the eigen functions, stands for the momentum vector and the spin quantum number; $u_{m\sigma}$ signifies the σ-component of the spinor eigen function u_m.

because it follows then with the help of (20.14 and 20.16):

$$[\psi_\sigma(x,t), \psi_{\sigma'}(x',t)] = [\pi_\sigma(x,t), \pi_{\sigma'}(x',t)] = 0,$$

$$[\pi_\sigma(x,t), \psi_{\sigma'}(x',t)] = \frac{h}{i}\sum_m u^*_{m\sigma}(x)\, u_{m\sigma'}(x') = \frac{h}{i}\,\delta_{\sigma\sigma'}\,\delta(x-x');$$

in this last equation we have used the fact that the eigen functions u_m form a complete system. One obtains for the energy $H = \int_V dx\, \mathsf{H}$ with the help of (20.13 to 20.16):

$$H = \sum_m E_m \cdot a^*_m\, a_m, \tag{20.17}$$

where $a^*_m\, a_m\ (\geqq 0)$, according to (6.18), represents the number N_m of electrons in the stationary state m. But it is known that one half of the eigen values E_m of the Dirac equation (20.15) are positive, and the other half negative $\left(E = \pm\, c\sqrt{(mc)^2 + p^2}\right)$, so that the energy (20.17) can assume both signs. (This fact does not depend on the order of the factors a^*_m, a_m.)

As mentioned before, a further objection is that the canonical quantization necessarily leads to Bose-Einstein statistics, whereas according to experimental evidence the electrons obey the Pauli exclusion principle, which means Fermi-Dirac statistics. This discrepancy with the experimental experiences is directly due to the fact that the occupation number $a^*_m\, a_m$ of an electron state is not limited from above, and is therefore not restricted to the values of 0 and 1. The reason for this is, of course, not due to the particular nature of the Hamiltonian (20.13). Any wave field with a Hamiltonian quadratic in the canonical variables q, p is equivalent to a system of harmonic oscillators with quantum numbers which have no upper limit. Since these quantum numbers here have the significance of occupation numbers of stationary states, one can see quite generally that the canonical commutation rules (1.7) violate the exclusion principle.

Jordan and Wigner have found a modification of the quantization method which takes account of the exclusion principle.[1] In order to formulate the respective equations for the special case of force-free electrons, we go back to the formulas (20.14), in which the u_m, as before, shall represent the eigen functions of the Dirac equation (20.15), whereas the definition of the operators a_m must be changed. We still assume that the operators a_m and a^*_m decrease

[1] *Z. Phys.* 47, 631, 1928.

or increase the particle number N_m by one, while they leave the remaining particle numbers $N_{m'}$ $(m' \neq m)$ unchanged [cf. (6.17)]:

$$(a_m)_{N'_1 \cdot \cdot, N''_1 \cdot \cdot} = (\overset{*}{a}_m)_{N''_1 \cdot \cdot, N'_1 \cdot \cdot} \sim \delta_{N'_m, N''_m - 1} \cdot \prod_{m' \neq m} \delta_{N'_{m'}, N''_{m'}} .$$

The numbers N_m, however, shall be restricted to the values 0 and 1, in accordance with the exclusion principle. The operators $a_m, \overset{*}{a}_m$ are thus, with respect to every particle number, two-row matrices, diagonal with respect to the numbers $N_{m'}$ $(m' \neq m)$, while they assume with respect to N_m the following form:

$$a_m = \eta_m \cdot \begin{pmatrix} 0 & 1 \\ 0 & 0 \end{pmatrix}, \quad \overset{*}{a}_m = \overset{*}{\eta}_m \cdot \begin{pmatrix} 0 & 0 \\ 1 & 0 \end{pmatrix}, \tag{20.18}$$

where the first row and column of the matrices refer always to the value $N_m = 0$, the second row and column refer to the value $N_m = 1$. The numerical factor η_m shall be determined later.[1] In other words: let F_{N_m} be a two-component function of the occupation number N_m, then we have for the functions $a_m F$ and $\overset{*}{a}_m F$ the following values:

$$(a_m F)_0 = \eta_m \cdot F_1, \qquad (a_m F)_1 = 0;$$
$$(\overset{*}{a}_m F)_0 = 0, \qquad (\overset{*}{a}_m F)_1 = \overset{*}{\eta}_m \cdot F_0 .$$

According to the rules of matrix multiplication one obtains from (20.18):

$$a_m a_m = \overset{*}{a}_m \overset{*}{a}_m = 0, \tag{20.19}$$

$$\begin{cases} \overset{*}{a}_m a_m = |\eta_m|^2 \cdot \begin{pmatrix} 0 & 0 \\ 0 & 1 \end{pmatrix} = |\eta_m|^2 \cdot N_m, \\ a_m \overset{*}{a}_m = |\eta_m|^2 \cdot \begin{pmatrix} 1 & 0 \\ 0 & 0 \end{pmatrix} = |\eta_m|^2 \cdot (1 - N_m); \end{cases} \tag{20.20}$$

Here the diagonal matrix with the eigen values N_m is called briefly N_m:

$$N_m = \begin{pmatrix} 0 & 0 \\ 0 & 1 \end{pmatrix}. \tag{20.21}$$

---

[1] η_m is assumed independent of N_m, so that the factor $\begin{pmatrix} 0 & 1 \\ 0 & 0 \end{pmatrix}$ in a_m commutes with η_m:

$$a_m = \eta_m \begin{pmatrix} 0 & 1 \\ 0 & 0 \end{pmatrix} = \begin{pmatrix} 0 & 1 \\ 0 & 0 \end{pmatrix} \eta_m .$$

The operators a_m and $\overset{*}{a}_m$ are then Hermitian conjugates.

If we require that $|\eta_m|^2 = 1$, it follows from (20.20):

$$a_m^* a_m + a_m a_m^* = 1; \tag{20.22}$$

The "anticommutator" of a_m and a_m^* is equal to the unit matrix, while the commutator $[a_m, a_m^*]$, according to (20.20), is equal to the diagonal matrix $(1 - 2N_m)$. It is seen that the commutators and anticommutators have interchanged their roles with respect to the commutation rules of a_m and a_m^* as compared with the canonical quantization ($[a_m, a_m^*] = 1$, $a_m a_m^* + a_m^* a_m = 1 + 2N_m$, cf. 6.18). Using the symbol:

$$[a, b]_+ = a\,b + b\,a$$

we sum up the formulas (20.19 and 20.22) as follows:

$$[a_m, a_m]_+ = [a_m^*, a_m^*]_+ = 0, \quad [a_m, a_m^*]_+ = 1.$$

As far as the commutation rules of two matrices with $m \neq m'$ are concerned, one might be tempted to assume them to commute with each other. This would, however, not result in simple commutation rules for ψ and π (20.14, 20.16). According to Jordan and Wigner, we replace, instead, the commutators by the corresponding anticommutators in all canonical commutation rules:

$$[a_m, a_{m'}]_+ = [a_m^*, a_{m'}^*]_+ = 0, \quad [a_m, a_{m'}^*]_+ = \delta_{mm'}. \tag{20.23}$$

Jordan and Wigner have proved that this quantized theory is equivalent with the theory in configuration space, in which the exclusion principle is taken into account by permitting only Schrödinger functions which are antisymmetrical in the coordinates of any two electrons.

It remains to be shown how the commutation rules (20.23) can be satisfied for the matrix pairs $m \neq m'$ with the assumption (20.18). In order to do this, one must assume a definite ordering of the stationary states $m = 1, 2, \ldots$ This ordering is, of course, arbitrary but must be maintained after it is once determined. We shall now set η_m equal to $+1$ or to -1 in (20.18), according to whether the number of occupied states with numbers $n < m$ is either even or odd.[1] This may be expressed by the following formula:

$$\eta_m = \prod_{n=1}^{m-1} (1 - 2N_n); \tag{20.24}$$

[1] But η_m shall not be dependent upon N_m; cf. footnote p. 173.

for in this product the occupied levels $n < m$ each contribute the factor $1 - 2N_n = -1$, while the unoccupied states contribute $1 - 2N_n = 1$. If we compare now the two matrices $a_m a_{m'}$ and $a_{m'} a_m$, defined by (20.18 and 20.24), where, for instance, $m < m'$, it is evident that the factor η_m has in both matrices the same sign, since the operator $a_{m'}$ does not change the occupation numbers N_n with $n < m$; the $\eta_{m'}$, on the contrary, have in both matrices opposite signs, since in the case of $a_{m'} a_m$ the previous application of the operators a_m had changed the occupation number N_m $(m < m')$ by 1, which is not true for $a_m a_{m'}$. Hence we have $a_{m'} a_m = - a_m a_{m'}$, in agreement with (20.23). The same is true if a_m is replaced by $\overset{*}{a}_m$ or a_m by $\overset{*}{a}_{m'}$. This reasoning can be expressed in formulas if one represents in η_m (20.24) the terms of the product $1 - 2N_n$ as matrices according to (20.21):

$$1 - 2N_n = \begin{pmatrix} 1 & 0 \\ 0 & -1 \end{pmatrix},$$

which is in agreement with $\eta_m \overset{*}{\eta}_m = \overset{*}{\eta}_m \eta_m = 1$. The products $\overset{(*)}{a}_m \overset{(*)}{a}_{m'}$ and $\overset{(*)}{a}_{m'} \overset{(*)}{a}_m$ then contain the same matrices with regard to all numbers N_n except N_m $(m < m')$, while the matrices with regard to N_m are equal and opposite to each other:

$$\begin{pmatrix} 0 & 1 \\ 0 & 0 \end{pmatrix}\begin{pmatrix} 1 & 0 \\ 0 & -1 \end{pmatrix} = -\begin{pmatrix} 1 & 0 \\ 0 & -1 \end{pmatrix}\begin{pmatrix} 0 & 1 \\ 0 & 0 \end{pmatrix} = -\begin{pmatrix} 0 & 1 \\ 0 & 0 \end{pmatrix},$$

$$\begin{pmatrix} 0 & 0 \\ 1 & 0 \end{pmatrix}\begin{pmatrix} 1 & 0 \\ 0 & -1 \end{pmatrix} = -\begin{pmatrix} 1 & 0 \\ 0 & -1 \end{pmatrix}\begin{pmatrix} 0 & 0 \\ 1 & 0 \end{pmatrix} = +\begin{pmatrix} 0 & 0 \\ 1 & 0 \end{pmatrix}.$$

With this, the existence of matrices with the commutation rules (20.23) is established for $m \neq m'$ as well as for $m = m'$. These relations, moreover, determine the operators a_m, $\overset{*}{a}_m$ uniquely, as Jordan and Wigner[1] have shown, if they are restricted to irreducible matrix systems and if one does not consider matrix systems as different from each other, which result from each other by similarity transformations $(a_m \rightarrow S^{-1} a_m S,\ \overset{*}{a}_m \rightarrow S^{-1} \overset{*}{a}_m S)$.

We substitute the so-defined operators a_m, $\overset{*}{a}_m$ into the expansions (20.14) and obtain thus the field operators ψ_σ immediately in a time-dependent form. They satisfy the Dirac equation (20.1). The connection between time-dependent and time-independent operators is established again by the general

[1] Loc. cit., p. 650 ff.

relationship (4.4) (with $E = H$), for it follows from (20.17 and 20.23):[1]

$$H\, a_m = a_m\, (H - E_m), \qquad H\, a^*_m = a^*_m\, (H + E_m),$$

i.e., the placing of $a^{(*)}_m$ to the left changes H into $H_{(\mp)}E_m$. Consequently one obtains by interchanging $a^{(*)}_m$ with $e^{\frac{it}{h}H}$:

$$e^{\frac{it}{h}H}\, a_m\, e^{-\frac{it}{h}H} = a_m\, e^{-\frac{it}{h}E_m}, \qquad e^{\frac{it}{h}H}\, a^*_m\, e^{-\frac{it}{h}H} = a^*_m\, e^{\frac{it}{h}E_m},$$

or according to (20.14):

$$\text{(20.25)} \quad \psi_\sigma(x, t) = e^{\frac{it}{h}H}\, \psi_\sigma(x, 0)\, e^{-\frac{it}{h}H}, \quad \psi^*_\sigma(x, t) = e^{\frac{it}{h}H}\, \psi^*_\sigma(x, 0)\, e^{-\frac{it}{h}H}$$

as was stipulated.

One obtains now for the operators ψ_σ (20.14), on account of (20.23), the following commutation rules:

$$\text{(20.26)} \qquad [\psi_\sigma(x, t), \psi_{\sigma'}(x', t')]_+ = [\psi^*_\sigma(x, t), \psi^*_{\sigma'}(x', t')]_+ = 0,$$

$$\text{(20.27)} \qquad [\psi_\sigma(x, t), \psi^*_{\sigma'}(x', t')]_+ = C_{\sigma\sigma'}(x - x', t - t'),$$

where:

$$\text{(20.28)} \qquad C_{\sigma\sigma'}(x - x', t - t') \equiv \sum_m e^{\frac{i}{h}(t' - t)E_m}\, u_{m\sigma}(x)\, u^*_{m\sigma'}(x').$$

It follows from the invariance under translations that the $C_{\sigma\sigma'}$ depend only on the coordinate differences $x - x'$. This is also confirmed by the following computation. For $t = t'$, in particular, the completeness relation for the orthogonal system of functions u_m yields:

$$\text{(20.29)} \qquad C_{\sigma\sigma'}(x - x', 0) \equiv \sum_m u_{m\sigma}(x)\, u^*_{m\sigma'}(x') = \delta_{\sigma\sigma'}\, \delta(x - x'),$$

so that one obtains the following commutation rule for the time-independent operators:

$$[\psi_\sigma(x, 0), \psi^*_{\sigma'}(x', 0)]_+ = \delta_{\sigma\sigma'}\, \delta(x - x'),$$

---

[1] For we have:

$$a^*_{m'}\, a_{m'} \cdot a_m = -a^*_{m'}\, a_m\, a_{m'} = \left(a_m\, a^*_{m'} - \delta_{mm'}\right) a_{m'}$$
$$= a_m \left(a^*_{m'}\, a_{m'} - \delta_{mm'}\right),$$
$$a^*_{m'}\, a_{m'} \cdot a^*_m = a^*_m \left(a^*_{m'}\, a_{m'} + \delta_{mm'}\right).$$

or, according to (20.12):

$$[\psi_\sigma(x, 0), \pi_{\sigma'}(x', 0)]_+ = -\frac{h}{i}\delta_{\sigma\sigma'}\delta(x - x'). \tag{20.30}$$

Here, again, the anticommutators take the place of the commutators. The calculation for $t \neq t'$, starting from (20.25 and 20.30), can be carried out in a similar way, as was done for the commutators in §4. We shall make the computation here in a slightly different (but essentially equivalent) way by making use of the Hamiltonian or the field equations, whereas in §4 only the Schrödinger-Gordon equation was used.

The factor $e^{\frac{i}{h}(t'-t)E_m} u_{m\sigma}(x)$ which appears in the terms in (20.28) can be considered as the σ-component of the spinor:

$$e^{\frac{i}{h}(t'-t)E_m} u_m(x) = e^{\frac{i}{h}(t'-t)\mathbf{E}} u_m(x) \tag{20.31}$$

where $\mathbf{E}$ is the operator defined by (20.1). The equation (20.31) holds because according to (20.15) $E_m u_m(x) = \mathbf{E}\, u_m(x)$ (one assumes that the exponential function is expanded into a power series). Hence, we can also write instead of (20.28):

$$C_{\sigma\sigma'}(x - x', t - t') = \sum_\varrho \left(e^{\frac{i}{h}(t'-t)\mathbf{E}}\right)_{\sigma\varrho} \cdot \sum_m u_{m\varrho}(x)\, u^*_{m\sigma'}(x'),$$

or, using (20.29):

$$C_{\sigma\sigma'}(x - x', t - t') = \left(e^{\frac{i}{h}(t'-t)\mathbf{E}}\right)_{\sigma\sigma'} \cdot \delta(x - x'), \tag{20.32}$$

where $\mathbf{E}$ acts as differential operator on the space coordinates x in the argument of the δ-function. As in §4, we assume that the latter is replaced by a regular function or is represented as a Fourier integral (4.24), so that $\mathbf{E}$ can be replaced under the integral by:

$$\mathbf{E}_k = c\, h\, (\alpha \cdot k) + m\, c^2 \beta$$

Interpreting $C_{\sigma\sigma'}$ as elements of a matrix C, we can abbreviate (20.32) in the form:

$$C(x, t) = e^{-\frac{i}{h}t\mathbf{E}} \cdot \delta(x). \tag{20.33}$$

We separate in the power series of the exponential function the even and odd terms in **E**:

$$e^{-\frac{i}{h}t\mathbf{E}} = \cos\left(\frac{t\mathbf{E}}{h}\right) - i\sin\left(\frac{t\mathbf{E}}{h}\right),$$

Taking into account that according to (20.1, 20.2)

$$\left(\frac{\mathbf{E}}{h}\right)^2 = c^2(\mu^2 - \nabla^2) \tag{20.34}$$

we can write for the cosine term, which contains only even powers of **E**:

$$\cos\left(\frac{t\mathbf{E}}{h}\right) = \cos\left(tc\sqrt{\mu^2 - \nabla^2}\right),$$

A corresponding equation holds for the sine term after factoring $\mathbf{E}/h$:

$$\sin\left(\frac{t\mathbf{E}}{h}\right) = \frac{\mathbf{E}}{h} \cdot \frac{\sin\left(tc\sqrt{\mu^2 - \nabla^2}\right)}{c\sqrt{\mu^2 - \nabla^2}}.$$

The functions of the operator $\sqrt{\mu^2 - \nabla^2}$ introduced here, contain as power expansions only integral powers of $\mu^2 - \nabla^2$ and ∇^2 is, when operating on e^{ikx}, equal to $-k^2$. From this there follows for C (20.33):

$$C(x,t) = \left\{\cos\left(tc\sqrt{\mu^2 - \nabla^2}\right) - \frac{i}{h}\mathbf{E}\frac{\sin\left(tc\sqrt{\mu^2 - \nabla^2}\right)}{c\sqrt{\mu^2 - \nabla^2}}\right\} \cdot \delta(x),$$

or, also:

$$C(x,t) = \left(\frac{\partial}{\partial t} - \frac{i}{h}\mathbf{E}\right) D(x,t), \tag{20.35}$$

where:

$$D(x,t) \equiv \frac{\sin\left(tc\sqrt{\mu^2 - \nabla^2}\right)}{c\sqrt{\mu^2 - \nabla^2}} \cdot \delta(x). \tag{20.36}$$

This D-function is nothing but the repeatedly used invariant D-function, as one sees immediately from the Fourier representation of $\delta(x)$ and $D(x,t)$ [cf. (4.24, .25)]. Thus the operator C is determined. Substitution of (20.35) into (20.27) yields the desired commutation rules:

$$\begin{aligned} [\psi_\sigma(x,t), \psi^*_{\sigma'}(x',t')]_+ &= \left(\delta_{\sigma\sigma'}\frac{\partial}{\partial t} - \frac{i}{h}\mathbf{E}_{\sigma\sigma'}\right) D(x-x', t-t') \\ &= \left\{\delta_{\sigma\sigma'}\frac{\partial}{\partial t} - c(\alpha_{\sigma\sigma'} \cdot \nabla) - \frac{imc^2}{h}\beta_{\sigma\sigma'}\right\} D(x-x', t-t'). \end{aligned} \tag{20.37}$$

In order to prove the Lorentz invariance of the commutation rules, it is convenient to introduce, instead of ψ^*, the "adjoint" wave function $\psi^\dagger = i\,\psi^*\,\beta$. If one multiplies (20.27) by $i\,\beta_{\sigma'\,\sigma''}$ and forms the sum over σ', one obtains for $[\psi_\sigma\,(x, t),\ \psi^\dagger_{\sigma''}\,(x', t')]_+$ the element (σ, σ'') of the matrix $(i\,C\,\beta)$, which in the denotations of (20.5) can be represented as follows:

$$i\,C\,(x, t)\,\beta = i\left(\frac{\partial}{\partial t} - \frac{i}{h}E\right)\beta\cdot D\,(x, t) = -c\left(\sum_\nu \gamma^{(\nu)}\frac{\partial}{\partial x_\nu} - \mu\right)D\,(x, t). \tag{20.38}$$

By including the remaining commutation rules after similar transformations, we have:

$$\begin{cases} [\psi_\sigma\,(x, t),\ \psi_{\sigma'}\,(x', t')]_+ = [\psi^\dagger_\sigma\,(x, t),\ \psi^\dagger_{\sigma'}\,(x', t')]_+ = 0, \\ [\psi_\sigma\,(x, t),\ \psi^\dagger_{\sigma'}\,(x', t')]_+ = -c\left\{\sum_\nu \gamma^{(\nu)}_{\sigma\sigma'}\frac{\partial}{\partial x_\nu} - \mu\,\delta_{\sigma\sigma'}\right\}D\,(x - x', t - t'). \end{cases} \tag{20.39}$$

Changing the reference system, one obtains for the transformed spinors $\bar{\psi} = S\,\psi$ and $\bar{\psi}^\dagger = \psi^\dagger\,S^{-1}$:

$$[\bar{\psi}_\varrho\,(x, t),\ \bar{\psi}^\dagger_{\varrho'}\,(x', t')]_+ = \sum_{\sigma\sigma'} S_{\varrho\sigma}\,(S^{-1})_{\sigma'\varrho'}\,[\psi_\sigma\,(x, t),\ \psi^\dagger_{\sigma'}\,(x', t')]_+$$

$$= -c\left\{\sum_\nu (S\gamma^{(\nu)}S^{-1})_{\varrho\varrho'}\frac{\partial}{\partial x_\nu} - \mu\,\delta_{\varrho\varrho'}\right\}D\,(x - x', t - t'),$$

This shows obviously, by virtue of (20.11), the invariance of the relations (20.39). The relativistic invariance of the Jordan-Wigner method of quantization is thus proven for the case of the force-free Dirac electron.

With regard to the current and energy momentum densities s_ν, $T_{\mu\nu}$, it should be said that the defining formulas (20.9, 20.10) together with the arrangement of factors, as chosen there, can be adopted for the quantized theory, since they satisfy all Hermitian conditions. The conservation laws $\sum_\nu \partial s_\nu/\partial x_\nu = 0$ and $\sum_\mu \partial T_{\mu\nu}/\partial x_\mu = 0$ are, of course, valid as they were in the unquantized theory, since the operators ψ (20.14), as spacetime functions, obey the same field equations (20.7) as the classical wave functions ψ.

Against this formulation of the theory one can still raise the objection that the energy $H = \sum_m E_m N_m$ [cf. (20.17, 20.20)] is not positive-definite. This deficiency can be removed, however, by a further modification of the formalism in the sense of Dirac's "theory of the positron" (theory of holes). The basis for this is laid by the quantization according to the exclusion principle.

Denoting briefly the energy levels of positive or negative energies ($E_m \gtrless 0$), positive or negative levels, respectively, we must interpret, according to Dirac, that state of the total system as "vacuum" in which all negative levels are occupied and all positive ones are empty:

$$N_m = \begin{cases} 1 & \text{for } E_m < 0, \\ 0 & \text{for } E_m > 0. \end{cases} \tag{20.40}$$

The definitions of the electric charge and the energy must be changed in such a way that their values for the vacuum state vanish. According to the old definitions (20.4) and (20.13 and 20.17) the total charge is:

$$\varepsilon h \int dx\, \psi^* \psi = \varepsilon h \sum_m a_m^* a_m = \varepsilon h \sum_m N_m$$

and the total energy:

$$\int dx\, \psi^* \boldsymbol{E} \psi = \sum_m E_m a_m^* a_m = \sum_m E_m N_m .$$

By subtracting from this the vacuum values:

$$\varepsilon h \sum_{\substack{m \\ (E_m < 0)}} 1 \qquad \text{and} \qquad \sum_{\substack{m \\ (E_m < 0)}} E_m$$

we obtain as new definitions:—for the charge:

$$e = \varepsilon h \sum_{\substack{m \\ (E_m > 0)}} N_m - \varepsilon h \sum_{\substack{m \\ (E_m < 0)}} (1 - N_m) = \varepsilon h \sum_{\substack{m \\ (E_m > 0)}} N_m - \varepsilon h \sum_{\substack{m \\ (E_m < 0)}} N'_m , \tag{20.41}$$

and for the energy:

$$H = \sum_{\substack{m \\ (E_m > 0)}} E_m N_m - \sum_{\substack{m \\ (E_m < 0)}} E_m (1 - N_m) = \sum_{\substack{m \\ (E_m > 0)}} |E_m| N_m + \sum_{\substack{m \\ (E_m < 0)}} |E_m| N'_m . \tag{20.42}$$

The diagonal matrix, which is introduced here:

$$N'_m = 1 - N_m = \begin{pmatrix} 1 & 0 \\ 0 & 0 \end{pmatrix} \tag{20.43}$$

[cf. (20.21)] when applied to the Schrödinger function yields zero if the level

m is occupied, and 1, if it is empty; in other words: N_m signifies the number of "holes" in the negative energy spectrum or the number of positrons. Such a hole ($N'_m = 1$) contributes, in fact, to the charge (20.41) the amount $-\epsilon h$ and to the energy (20.42) the amount $|E_m|$, while in the positive spectrum ($N_m = 1$) the charge $+\epsilon h$ and the energy $|E_m|$ are assigned to an existing electron. With this artifice, Dirac succeeded in making the energy positive-definite, whereas the charge has lost its definite character, as should be in view of the existence of the positrons.

In order to discuss also the momentum and the angular momentum from this point of view, we notice that the expression for the momentum density ($\mathbf{G}_k = T_{4k}/i\,c$), given by (20.10):

$$\mathbf{G} = \frac{h}{4i}\left\{\psi^*\left(\nabla\psi - \alpha\frac{1}{c}\frac{\partial\psi}{\partial t}\right) - \left(\nabla\psi^* - \frac{1}{c}\frac{\partial\psi^*}{\partial t}\alpha\right)\psi\right\}$$

can be transformed with the help of the Dirac equation (20.1) and the α-commutation rules (20.2) as follows:

$$\mathbf{G} = \frac{h}{2i}\{\psi^*(\nabla\psi) - (\nabla\psi^*)\psi\} + \frac{h}{4}\nabla\times(\psi^*\sigma\psi); \qquad (20.44)$$

σ denotes here the matrix vector with the components $\sigma^{(1)} = -i\,\alpha^{(2)}\alpha^{(3)}$, . . ., (cyclic). The term $\nabla\times\psi^*\sigma\psi$ does not contribute anything to the total momentum $G = \int_V dx\,\mathbf{G}$. We substitute the expansions of the eigen functions (20.14) into the remaining terms and take into account that $u_m \sim e^{ik_m x}$, hence $\nabla u_m = i\,k_m\,u_m$. On account of the orthogonality relations (20.15), it follows for the total momentum:

$$G = \sum_m h\,k_m\,a^*_m\,a_m = \sum_m h\,k_m\,N_m.$$

If we again subtract from it the vacuum value:

$$\sum_{\substack{m \\ (E_m<0)}} h\,k_m,$$

we obtain the corrected value for the momentum:

$$G = \sum_{\substack{m \\ (E_m>0)}} h\,k_m\,N_m - \sum_{\substack{m \\ (E_m<0)}} h\,k_m\,N'_m. \qquad (20.45)$$

According to this, an occupied single state m in the positive energy spectrum has the momentum $+h\,k_m$, an unoccupied state m in the negative spectrum (i.e., the respective positron), the momentum $-\,h\,k_m$.

We decompose the angular momentum $M = \int dx\; x \times \mathsf{G}$ [analogous to the formulas (12.56–12.60) and (16.50)] into two terms:

$$M = M^0 + M',$$

$$M^0 = \frac{h}{2i}\int\limits_V dx\; x \times \{\psi^* \nabla \psi - \nabla \psi^* \psi\},$$

$$M' = \frac{h}{4}\int\limits_V dx\; x \times (\nabla \times \psi^* \sigma \psi) = \frac{h}{2}\int\limits_V dx\; \psi^* \sigma \psi.$$

Applied to particles of non-relativistic velocity the term M^0, which is independent of the orientation of the spin, yields the orbital angular momentum, the term M', the spin angular momentum, which is of interest here. Since $u_m \sim e^{ik_m x}$ and σ is independent of x, M' is a sum of terms arising from the contributions of the wave functions belonging to the different k-values:

$$M' = \sum_k M'_{(k)},$$

As in the case of the meson with spin 1, we can restrict our discussion to the term $M'_{(0)}$ (particle at rest). To the momentum value $k = 0$ there belong four eigen functions $u_1 \ldots u_4$ with the energy eigen values:

$$E_1 = E_2 = +\,m\,c^2, \qquad E_3 = E_4 = -\,m\,c^2;$$

Admitting a unitary transformation the functions $u_1 \ldots u_4$ can be chosen in such a way that we obtain for a component of the vector matrix $\int dx\; u_m^* \sigma\, u_{m'}$ $(= V \cdot u_u^* \sigma\, u_{m'})$, for instance, for the x_1-component, the following relations:

$$\int dx\; u_m^* \sigma^{(1)} u_{m'} = 0 \quad \text{for } m \neq m' \; (m, m' = 1 \ldots 4),$$

$$\int dx\; u_m^* \sigma^{(1)} u_m = \begin{cases} +1 & \text{for } m = 1 \text{ and } m = 3, \\ -1 & \text{for } m = 2 \text{ and } m = 4. \end{cases}$$

Hence, it follows for the x_1-component of $M'_{(0)}$:

$$M'_1 = \frac{h}{2}\sum_{m=1}^{4}\sum_{m'=1}^{4} a_m^* a_{m'}\, e^{\frac{it}{h}(E_m - E_{m'})} \cdot \int dx\; u_m^* \sigma^{(1)} u_{m'}$$

$$= \frac{h}{2}(N_1 - N_2 + N_3 - N_4).$$

In "vacuo" the states $m = 3$ and $m = 4$ ($E_m < 0$) are occupied. The subtraction of the vacuum value from M_1' transforms thus N_3 into $N_3 - 1 = - N_3'$ and N_4 into $N_4 - 1 = - N_4'$, so that the corrected value of M_1' is:

$$M_1' = \frac{h}{2}(N_1 - N_2 - N_3' + N_4'); \tag{20.46}$$

i.e., N_1 electrons (at rest) and N_4' positrons have the spin components $+ h/2$, N_2 electrons and N_3' positrons have the spin components $- h/2$ in the direction x_1. Thus the physical meaning of the quantum numbers N_m and N_m' is completely established.

The state of the total system which is characterized by the values of all quantum numbers N_m, N_m' must evidently be counted as a single state. This, together with the restriction of occupation numbers to the values 0 and 1, corresponds to the statistical weight for the case of the Fermi-Dirac statistics.

The subtraction rule as formulated so far is not quite unambiguous, since we are dealing with subtractions of infinite expressions (divergent sums). We solved this difficulty for the above evaluation of charge, energy, etc., by carrying out the subtraction for each single term m of the sum. This method, however, does not always allow a generalization (electrons in force fields, cf. §21). In order to remove this deficiency, we introduce the following "density matrix":[1]

$$r_{\sigma\sigma'}(x, t; x', t') = \frac{1}{2}\sum_{m,m'}\left(a_{m'}^* a_m - a_m a_{m'}^* + \frac{E_m}{|E_m|}\delta_{mm'}\right) \cdot e^{\frac{i}{h}(t' E_{m'} - t E_m)} u_{m\sigma}(x)\, u_{m'\sigma'}^{\dagger}(x'), \tag{20.47}$$

and replace the former definitions for the charge- and energy-momentum quantities (20.9, 20.10) by the following definitions:

$$s_\nu = \lim_{\substack{t'=t\\ x'=x}} \varepsilon h c \sum_{\sigma,\sigma'} \gamma_{\sigma'\sigma}^{(\nu)} r_{\sigma\sigma'}(x, t; x', t'), \tag{20.48}$$

$$T_{\mu\nu} = \lim_{\substack{t'=t\\ x'=x}} \frac{hc}{4i} \sum_{\sigma,\sigma'} \left\{\gamma_{\sigma'\sigma}^{(\mu)}\left(\frac{\partial}{\partial x_\nu} - \frac{\partial}{\partial x_\nu'}\right) + \gamma_{\sigma'\sigma}^{(\nu)}\left(\frac{\partial}{\partial x_\mu} - \frac{\partial}{\partial x_\mu'}\right)\right\} \cdot r_{\sigma\sigma'}(x, t; x', t'). \tag{20.49}$$

In order to examine their agreement with the primitive subtraction rule, we shall compare the new with the old definitions. For instance, with regard to

[1] Fock, *C. R. Leningrad 1933*, 267; Furry and Oppenheimer, *Phys. Rev. 45*, 245, 1934; Peierls, *Proc. Roy. Soc. London 146*, 420, 1934; Dirac, *Proc. Cambridge Phil. Soc. 30*, 150, 1934; Heisenberg, *Z. Phys. 90*, 209, 1934.

the 4-current s_ν, we find, for the difference of the expressions (20.48) and (20.9) with the help of (20.14 and 20.23):

$$\varepsilon h c \frac{1}{2} \sum_m \left(-1 + \frac{E_m}{|E_m|}\right) u_m^\dagger \gamma^{(\nu)} u_m = -\varepsilon h c \sum_{\substack{m \\ (E_m < 0)}} u_m^\dagger \gamma^{(\nu)} u_m, \tag{20.50}$$

which is equal to the negative part of the expectation value of the current (20.9) for vacuum. A similar identity holds for the energy-momentum tensor. From this it is easily seen that one obtains for the integral quantities e, H, G, M' based on (20.47, 20.48, 20.49) in fact, the same values as with the original subtraction rule [cf. (20.41, 20.42, 20.45, 20.46)]. But the new definitions (20.47, 20.48, 20.49) are, in contrast to the earlier ones, free from diverging sums of the type (20.50); for if we consider the diagonal elements of the matrix (20.47), i.e., the terms $m = m'$ (only these are of interest):

$$\frac{1}{2} \sum_m \left(a_m^* a_m - a_m a_m^* + \frac{E_m}{|E_m|}\right) e^{\frac{i}{\hbar}(t'-t) E_m} u_{m\sigma}(x) u_{m\sigma'}^\dagger(x')$$

$$= \left\{ \sum_{\substack{m \\ (E_m > 0)}} N_m - \sum_{\substack{m \\ (E_m < 0)}} N'_m \right\} e^{\frac{i}{\hbar}(t'-t) E_m} u_{m\sigma}(x) u_{m\sigma'}^\dagger(x'),$$

we obtain only finite sums provided that only a finite number of positive states are occupied and only a finite number of negative states are unoccupied, i.e., that only a finite number of particles are present.

We decompose r (20.47) into two sums:

$$r = R + S, \tag{20.51}$$

$$R_{\sigma\sigma'}(x, t; x', t') = \frac{1}{2} \sum_{m, m'} \left(a_{m'}^* a_m - a_m a_{m'}^*\right) e^{\frac{i}{\hbar}(t' E_{m'} - t E_m)} u_{m\sigma}(x) u_{m'\sigma'}^\dagger(x') \tag{20.52}$$

$$S_{\sigma\sigma'}(x, t; x', t') = \frac{1}{2} \sum_m \frac{E_m}{|E_m|} e^{\frac{i}{\hbar}(t'-t) E_m} u_{m\sigma}(x) u_{m\sigma'}^\dagger(x'), \tag{20.53}$$

Both sums converge separately, provided that the world vector $(x - x', t - t')$ is not a zero vector $(|x - x'|^2 - c^2 (t - t')^2 \neq 0)$. The limiting process $t' \to t$, $x' \to x$, considered in (20.48, 20.49), can, however, only be carried out if R and S are combined to r. R can also be expressed according to (20.14):

$$R_{\sigma\sigma'}(x, t; x', t') = \frac{1}{2} \{\psi_{\sigma'}^\dagger(x', t') \psi_\sigma(x, t) - \psi_\sigma(x, t) \psi_{\sigma'}^\dagger(x', t')\}. \tag{20.54}$$

S on the other hand is a function of $x - x'$ and $t - t'$, which we can calculate as follows: if we set in (20.53) formally [similarly as in (20.31), cf. also (20.34)]:

$$\frac{1}{|E_m|} u_m(x) = \frac{1}{\sqrt[+]{E^2}} u_m(x) = \frac{1}{hc\sqrt[+]{\mu^2 - \nabla^2}} u_m(x),$$

where ∇^2 again signifies $-k^2$ $(u_m \sim e^{ikx})$, then the matrix S (with the elements $S_{\sigma\sigma'}$) can be represented by the matrix C (20.28) as follows:

$$S(x, t; x', t') = \frac{1}{2}\left(-\frac{h}{i}\frac{\partial}{\partial t}\right)\frac{1}{hc\sqrt[+]{\mu^2 - \nabla^2}} \cdot iC(x - x', t - t')\beta$$

(note that $u_m^\dagger = i u_m^* \beta$). The matrix S can thus be expressed in terms of the invariant D-function according to (20.35 or 20.38):

$$S(x, t; x', t') = -\frac{i}{2} c\left(\sum_\nu \gamma^{(\nu)} \frac{\partial}{\partial x_\nu} - \mu\right) D'(x - x', t - t'), \tag{20.55}$$

where:

$$D'(x, t) = \frac{1}{c}\frac{\partial}{\partial t}\frac{1}{\sqrt[+]{\mu^2 - \nabla^2}} D(x, t).$$

In these formulas, C and D are to be considered as Fourier series, so that the operator $1/\sqrt{\mu^2 - \nabla^2}$, under the integral, if applied to e^{ikx}, assumes the value $1/\sqrt{\mu^2 + k^2}$. From the Fourier representation of D (4.25) there follows that of D':

$$D'(x, t) = \frac{1}{(2\pi)^3}\int dk\, e^{ikx} \frac{\cos\left(tc\sqrt{\mu^2 + k^2}\right)}{c\sqrt[+]{\mu^2 + k^2}}. \tag{20.56}$$

D' is obviously a Lorentz-invariant function in the same sense as D; for D' is determined by the sum of the two invariant functions (4.27), while D is determined by their difference. By carrying out the k-space integration in (20.56), D' can be expressed in terms of Hankel's cylinder functions with the argument $\mu\sqrt{c^2 t^2 - x^2}$. We shall not discuss this further.[1]

[1] Cf. the papers quoted in footnote 1, p. 25; for $\mu = 0$:

$$D'(x, t) = \frac{1}{(2\pi)^3}\int dk \cos kx \frac{\cos |k| ct}{c|k|} = \frac{1}{2\pi^2 c}\frac{1}{x^2 - c^2 t^2};$$

cf. also (19.50, 19.51).

The fundamental equations of the theory can now be stated without referring to the expansions in eigen functions (20.14): the properties of the operators $\psi_\sigma(x, t)$, $\psi^*_\sigma(x, t)$ are determined by the Dirac equation (20.1 and 20.7) and by the commutation rules (20.26, 20.37, or 20.39); the physical interpretation is based on the defining formulas (20.48, 20.49) for s_ν, $T_{\mu\nu}$, in conjunction with (20.51, 20.54, 20.55, 20.56).

While the primitive subtraction method distinguishes the electrons rather than the positrons, by interpreting the latter as electron holes, the formulation with the density matrix r has the additional advantage that it is symmetrical with respect to the two kinds of particles of opposite charge. We verify this by defining first:

$$a^*_m = a'_m, \quad a_m = a'^*_m. \tag{20.57}$$

This transformation of the operators a leaves the commutation rules (20.23) obviously invariant:

$$[a'_m, a'_{m'}]_+ = [a'^*_m, a'^*_{m'}]_+ = 0, \quad [a'_m, a'^*_{m'}]_+ = \delta_{mm'}.$$

It corresponds to an exchange of occupied and unoccupied states, i.e., of the numbers N_m and $N'_m = 1 - N_m$; for while the operator a_m decreases the number N_m by 1, a'_m decreases the number N'_m by 1. Considering now that N_m signifies an electron number for a positive level, N'_m a positron number for a negative level, it becomes evident that a transformation by which electrons and positrons are to be interchanged must also imply a reflection of the energy scale. Such a transformation which also includes the transformation (20.57) is given by:

$$\psi'_\sigma = \psi^\dagger_\sigma. \tag{20.58}$$

The spinor function ψ' thus defined does not satisfy the same Dirac equation (20.7) as ψ, for $\psi^\dagger$ satisfies:

$$\sum_\nu \frac{\partial \psi^\dagger}{\partial x_\nu} \gamma^{(\nu)} - \mu \psi^\dagger = 0$$

and it follows for ψ':

$$\left(\sum_\nu \gamma'^{(\nu)} \frac{\partial}{\partial x_\nu} + \mu\right) \psi' = 0, \tag{20.59}$$

where $\gamma'^{(\nu)}$ stands for the "transposed matrix" $\gamma^{(\nu)}$ with the negative sign:

$$\gamma'^{(\nu)}_{\varrho\sigma} = -\gamma^{(\nu)}_{\sigma\varrho}. \tag{20.60}$$

But since these Hermitian matrices $\gamma'^{(\nu)}$ have the same commutation rules as the $\gamma^{(\nu)}$ ($\gamma'^{(\nu)}\gamma'^{(\mu)} + \gamma'^{(\mu)}\gamma'^{(\nu)} = 2\,\delta_{\mu\nu}$), the equation (20.59) can, by a suitable transformation $\psi' = S\,\psi''$, with $\gamma'^{(\nu)} S = S\,\gamma^{(\nu)}$, $SS^* = S^*S = 1$, be brought into the form (20.7)[1] and hence can be assumed to be equivalent with (20.7). If one defines, furthermore, the adjoint function to ψ' according to (20.6) as $\psi'^\dagger = i\,\psi'^*\,\gamma'^{(4)}$, it follows, according to (20.60), that $\psi'^\dagger = -\,i\,\gamma^{(4)}\,\psi'^*$, and since, according to (20.58) $\psi'^* = (i\,\psi^*\,\gamma^{(4)})^* = -\,i\,\gamma^{(4)}\,\psi$:

$$\psi_\sigma'^\dagger = -\psi_\sigma. \tag{20.61}$$

One obtains for ψ' and $\psi'^\dagger$ according to (20.58, 20.61, and 20.39) the following commutation rules:

$$[\psi_\sigma'(x,t), \psi_{\sigma'}'^\dagger(x',t')]_+ = c\left\{\sum_\nu \gamma_{\sigma'\sigma}^{(\nu)}\frac{\partial}{\partial x_\nu'} - \mu\,\delta_{\sigma'\sigma}\right\} D(x'-x, t'-t),$$

and if one considers (20.60) and (4.29):

$$[\psi_\sigma'(x,t), \psi_{\sigma'}'^\dagger(x',t')]_+ = -c\left\{\sum_\nu \gamma_{\sigma\sigma'}'^{(\nu)}\frac{\partial}{\partial x_\nu} - \mu\,\delta_{\sigma\sigma'}\right\} D(x-x', t-t'),$$

while, of course:

$$[\psi_\sigma'(x,t), \psi_{\sigma'}'(x',t')]_+ = [\psi_\sigma'^\dagger(x,t), \psi_{\sigma'}'^\dagger(x',t')]_+ = 0.$$

Thus we have shown that the transformation (20.58) leaves the commutation rules (20.39) invariant. In order to see how the quantities s_ν, $T_{\mu\nu}$ transform, we shall express the density matrix r in terms of ψ' and $\psi'^\dagger$. For R (20.54) we find:

$$\begin{aligned} R_{\sigma\sigma'}(x,t;x',t') &= \frac{1}{2}\{\psi_\sigma'^\dagger(x,t)\,\psi_{\sigma'}'(x',t') - \psi_{\sigma'}'(x',t')\,\psi_\sigma'^\dagger(x,t)\} \\ &= R_{\sigma'\sigma}'(x',t';x,t). \end{aligned}$$

Considering (20.60) and using the symmetry property $D'(x,t) = D'(-x,-t)$, we may write for S (20.55):

$$\begin{aligned} S_{\sigma\sigma'}(x,t;x',t') &= -\frac{i}{2}c\left(\sum_\nu \gamma_{\sigma'\sigma}'^{(\nu)}\frac{\partial}{\partial x_\nu'} - \mu\,\delta_{\sigma'\sigma}\right) D'(x'-x, t'-t) \\ &= S_{\sigma'\sigma}'(x',t';x,t). \end{aligned}$$

The transformation (20.58) thus introduces in R, as well as in S, and conse-

[1] Cf., for instance, Pauli, *Ann. inst. Henri Poincaré* 6, 109, 1936, especially §4.

quently also in $r = R + S$, an interchange of the variables x, t, σ with the primed variables x', t', σ':

$$r_{\sigma\sigma'}(x, t; x', t') = r'_{\sigma'\sigma}(x', t'; x, t).$$

Substituting this into the formulas (20.48, 20.49) for s_ν, $T_{\mu\nu}$, we may there interchange the primed and non-primed variables, since this is only a change of notation. Applying (20.60) once more we obtain:

$$s_\nu = \lim_{\substack{t'=t \\ x'=x}} (-\varepsilon)\, h\, c \sum_{\sigma, \sigma'} \gamma'^{(\nu)}_{\sigma'\sigma}\, r'_{\sigma\sigma'}(x, t; x', t') = -s'_\nu,$$

$$T_{\mu\nu} = \lim_{\substack{t'=t \\ x'=x}} \frac{hc}{4i} \sum_{\sigma, \sigma'} \left\{ \gamma'^{(\mu)}_{\sigma'\sigma} \left(\frac{\partial}{\partial x_\nu} - \frac{\partial}{\partial x'_\nu} \right) + \gamma'^{(\nu)}_{\sigma'\sigma} \left(\frac{\partial}{\partial x_\mu} - \frac{\partial}{\partial x'_\mu} \right) \right\}.$$

$$\cdot\, r'_{\sigma\sigma'}(x, t; x', t') = + T'_{\mu\nu}.$$

Hence the result is that the energy-momentum tensor is invariant under the transformation discussed, while the 4-current s_ν changes its sign, according to the fact that ϵh stands for the electronic charge in s_ν, whereas it stands for the positron charge in s'_ν. It is gratifying that the formalism permits this exchange of electrons and positrons, not only in view of the experimental facts, but also because it stresses the analogy of the theory for particles with spin $\frac{1}{2}$ to the earlier discussed theories of particles with integral spin, where the symmetry with respect to positively- and negatively-charged particles is evident. (cf. §§8 and 12).

§ 21. Electrons in the Electromagnetic Field

With the notation (13.2):

$$\partial_\nu \equiv \frac{\partial}{\partial x_\nu} - \frac{i\varepsilon}{c}\Phi_\nu, \qquad \partial^*_\nu \equiv \frac{\partial}{\partial x_\nu} + \frac{i\varepsilon}{c}\Phi_\nu, \tag{21.1}$$

where the Φ_ν represents the given electromagnetic potentials, we obtain the Lagrangian of the electrons in the field by replacing $\partial/\partial x_\nu$ by ∂_ν (similarly as in §§11 and 13) in the Lagrangian for the field-free electrons (20.8):

$$L = -\frac{hc}{i}\psi^\dagger \left(\sum_\nu \gamma^{(\nu)} \partial_\nu + \mu \right) \psi. \tag{21.2}$$

From this there follow readily these field equations:

$$\sum_\nu \gamma^{(\nu)} \partial_\nu \psi + \mu\psi = 0, \qquad \sum_\nu \partial^*_\nu \psi^\dagger \gamma^{(\nu)} - \mu\psi^\dagger = 0. \tag{21.3}$$

The 4-current (without subtraction terms) remains the same as in the field-free case:

$$s_\nu = \varepsilon h c \cdot \psi^\dagger \gamma^{(\nu)} \psi, \tag{21.4}$$

while the energy-momentum tensor is obtained from (20.10) by substitution of $\partial_\nu \psi$ for $\partial\psi/\partial x_\nu$ and of $\partial_\nu^* \psi^\dagger$ for $\partial\psi^\dagger/\partial x_\nu$:

$$T_{\mu\nu} = \frac{h c}{4 i} \{\psi^\dagger \gamma^{(\mu)} \partial_\nu \psi + \psi^\dagger \gamma^{(\nu)} \partial_\mu \psi - (\partial_\nu^* \psi^\dagger) \gamma^{(\mu)} \psi - (\partial_\mu^* \psi^\dagger) \gamma^{(\nu)} \psi\}. \tag{21.5}$$

With the help of (21.3, 21.4, 21.5) the following conservation laws are found:[1]

$$\sum_\nu \frac{\partial s_\nu}{\partial x_\nu} = 0, \qquad \sum_\mu \frac{\partial T_{\mu\nu}}{\partial x_\mu} = -\frac{1}{c} \sum_\mu s_\mu F_{\mu\nu}. \tag{21.6}$$

Under the gauge transformation [cf. (11.4)]:

$$\Phi_\nu \to \Phi_\nu + \frac{\partial \Lambda}{\partial x_\nu}, \qquad \psi \to e^{\frac{i\varepsilon}{c}\Lambda} \psi, \qquad \psi^\dagger \to \psi^\dagger e^{-\frac{i\varepsilon}{c}\Lambda} \tag{21.7}$$

L, s_ν, $T_{\mu\nu}$ as well as the field equations (21.3) are invariant. With:

$$\pi_\sigma = \frac{\partial L}{\partial \dot{\psi}_\sigma} = h (\psi^\dagger \gamma^{(4)})_\sigma = i h \psi_\sigma^*, \quad \pi_\sigma^\dagger = \frac{\partial L}{\partial \dot{\psi}_\sigma^\dagger} = 0 \tag{21.8}$$

one finds for the Hamiltonian:

$$\begin{aligned} H &= h \psi^\dagger \gamma^{(4)} \dot{\psi} - L = i h c \psi^\dagger \gamma^{(4)} \frac{\partial \psi}{\partial x_4} - L \\ &= -c \pi \sum_{k=1}^{3} \alpha^{(k)} \partial_k \psi - i \mu c \pi \beta \psi - i \varepsilon \pi \psi \cdot \Phi_0 \end{aligned} \tag{21.9}$$

($\Phi_4 = i \Phi_0$). One can easily verify that $H + i \epsilon \pi \psi \Phi_0$ agrees with $-T_{44}$ except for terms which either vanish according to the Dirac equations (21.3), or are expressible as spatial divergences; i.e., H differs from $-T_{44}$ essentially by the term $-i \epsilon \pi \psi \Phi_0 = \rho \Phi_0$.

With regard to the quantization of the theory, we remind the reader that in all earlier cases, with quantization according to the Bose-Einstein statistics, the commutation rules for the time-independent operators ψ_σ, π_σ were the same as for the force-free case; this is inherent in the nature of the canonical quantization. Hence we shall try also here, where we quantize according to the exclusion principle, the same commutation rules for the time-independent

[1] Cf., for instance, Pauli, *Handbuch der Physik*, Geiger-Scheel, Bd. 24, I. Teil, p. 235.

operators ψ_σ, π_σ as in the theory of the force-free electrons [cf. (20.26, 20.30)]:

$$[\psi_\sigma(x), \psi_{\sigma'}(x')]_+ = [\pi_\sigma(x), \pi_{\sigma'}(x')]_+ = 0, \; [\psi_\sigma(x), \pi_{\sigma'}(x')]_+ = i h \, \delta_{\sigma\sigma'} \, \delta(x - x'). \tag{21.10}$$

One may satisfy these relations by expanding with respect to any complete orthogonal system (for instance, the eigen functions of the force-free electrons):

$$\psi_\sigma(x) = \sum_m a_m \, u_{m\sigma}(x), \qquad \pi_\sigma(x) = i h \sum_m \overset{*}{a}_m \, \overset{*}{u}_{m\sigma}(x), \tag{21.11}$$

where the coefficients a_m, $\overset{*}{a}_m$ are matrices of the type (20.18) with the commutation rules (20.23); this assumption obviously is in agreement with the postulate (21.10). On account of (21.9 and 21.10) one finds for $\dot{\psi}(x) = \frac{i}{h}[H, \psi(x)]$ and $\dot{\pi}(x) = \frac{i}{h}[H, \pi(x)]$:[1]

$$\begin{cases} \dot{\psi} = -c \sum_k \alpha^{(k)} \, \partial_k \psi - i \mu c \beta \psi - i \varepsilon \Phi_0 \psi, \\ \dot{\pi} = -c \sum_k \overset{*}{\partial}_k \pi \, \alpha^{(k)} + i \mu c \pi \beta + i \varepsilon \Phi_0 \pi, \end{cases} \tag{21.12}$$

which corresponds to the field equations (21.3) [cf. (21.8)].

In order to study the relativistic invariance of the theory we introduce time-dependent operators $\psi(x, t)$, $\pi(x, t)$ in such a way that they satisfy the differential equations (21.12). We can assume them to be represented in the form of a Taylor expansion with respect to powers of t:

---

[1] If we write H in the form:

$$H = \sum_{\varrho, \sigma} \pi_\varrho \, O_{\varrho\sigma} \, \psi_\sigma,$$

where $O_{\rho\sigma}$ is a differential operator acting on ψ_σ, it follows, for instance:

$$\begin{aligned} H \psi_\tau(x) &= -\int dx' \sum_{\varrho, \sigma} \pi_\varrho(x') \, \psi_\tau(x) \, O_{\varrho\sigma} \, \psi_\sigma(x') \\ &= -\int dx' \sum_{\varrho, \sigma} \{-\psi_\tau(x) \, \pi_\varrho(x') + i h \, \delta_{\varrho\tau} \, \delta(x - x')\} \, O_{\varrho\sigma} \, \psi_\sigma(x') \\ &= \psi_\tau(x) H - i h \sum_\sigma O_{\tau\sigma} \, \psi_\sigma(x), \end{aligned}$$

hence:

$$\dot{\psi}_\tau = \sum_\sigma O_{\tau\sigma} \, \psi_\sigma = \frac{\partial H}{\partial \pi_\tau}.$$

$$\psi(x,t) = \psi(x) + t\,\dot{\psi}(x) + \frac{1}{2}\,t^2\,\ddot{\psi}(x) + \ldots,$$

$$\pi(x,t) = \pi(x) + t\,\dot{\pi}(x) + \frac{1}{2}\,t^2\,\ddot{\pi}(x) + \ldots.$$

If we substitute these developments into the anticommutators $[\psi_\sigma(x,t), \psi_{\sigma'}(x',t')]_+$ etc., we obtain, with (21.10 and 21.12):[1]

(21.13)
$$[\psi_\sigma(x,t), \psi_{\sigma'}(x',t')]_+ = [\pi_\sigma(x,t), \pi_{\sigma'}(x',t')]_+ = 0,$$

$$[\psi_\sigma(x,t), \pi_{\sigma'}(x',t')]_+ = i\,h\Big\{\delta_{\sigma\sigma'} +$$
$$+\, t\Big(-c\sum_k \alpha^{(k)}_{\sigma\sigma'}\Big\{\frac{\partial}{\partial x_k} - \frac{i\,\varepsilon}{c}\,\Phi_k(x,0)\Big\} - i\,\mu\,c\,\beta_{\sigma\sigma'} - i\,\varepsilon\,\delta_{\sigma\sigma'}\,\Phi_0(x,0)\Big)$$
$$+\, t'\Big(-c\sum_k \alpha^{(k)}_{\sigma\sigma'}\Big\{\frac{\partial}{\partial x'_k} + \frac{i\,\varepsilon}{c}\,\Phi_k(x',0)\Big\} + i\,\mu\,c\,\beta_{\sigma\sigma'} + i\,\varepsilon\,\delta_{\sigma\sigma'}\,\Phi_0(x',0)\Big)$$
$$+ \ldots\Big\}\,\delta(x-x').^{2}$$

Here those parts of the Taylor expansions which are independent of Φ_ν can be summed up and, indeed, the result may be taken from the theory of the field-free electron, i.e., from formula (20.37):

(21.14)
$$[\psi_\sigma(x,t), \pi_{\sigma'}(x',t')]_+ =$$
$$= i\,h\Big\{\delta_{\sigma\sigma'}\frac{\partial}{\partial t} - c\,(\alpha_{\sigma\sigma'}\cdot\nabla) - i\,\mu\,c\,\beta_{\sigma\sigma'}\Big\}\,D(x-x', t-t')$$
$$+ \{\varepsilon\,h\,(t'-t)\big[(\alpha_{\sigma\sigma'}\cdot\Phi(x,0)) - \delta_{\sigma\sigma'}\,\Phi_0(x,0)\big] + \ldots\}\,\delta(x-x')$$

(here we used: $\Phi_\nu(x',0)\,\delta(x-x') = \Phi_\nu(x,0)\,\delta(x-x')$, which evidently is permitted).

If we now carry out a Lorentz transformation and if we consider two world

[1] For $t' = t$, of course, we must have:

$$\big[\psi_\sigma(x,t), \pi_{\sigma'}(x',t)\big]_+ = i\,h\,\delta_{\sigma\sigma'}\,\delta(x-x'),$$

for any t-values. This is, in fact, compatible with (21.12):

$$\frac{\partial}{\partial t}\big[\psi_\sigma(x,t), \pi_{\sigma'}(x',t)\big]_+$$
$$= \big[\dot{\psi}_\sigma(x,t), \pi_{\sigma'}(x',t)\big]_+ + \big[\psi_\sigma(x,t), \dot{\pi}_{\sigma'}(x',t)\big]_+ = 0.$$

The time origin is thus not distinguished. This means that without loss in generality one may develop in powers of t or t' instead of $t' - t$ (one could, for instance, choose $t = 0$).

points x, t and x', t', which are simultaneous in the new reference system then we have:

$$t' - t = \sum_{k=1}^{3} a_k (x'_k - x_k), \tag{21.15}$$

The term linear in $t' - t$ vanishes in (21.14), since $(x'_k - x_k)\, \delta(x - x') = 0$. If we restrict the discussions to infinitesimal Lorentz transformations, so that terms of higher order in $t' - t$ can be neglected, the equation (21.14), assuming (21.15), is equal to the corresponding equation for the field-free case. Since the Lorentz invariance of the commutation rules in the field-free case has already been proved (§20), we can deduce for the transformed operators $\bar{\psi} = S\psi$, $\bar{\pi} = \pi S^* = \pi \beta S^{-1} \beta$:

$$\left[\bar{\psi}_\sigma(\bar{x}, \bar{t}), \bar{\pi}_{\sigma'}(\bar{x}', \bar{t})\right]_+ = i\hbar\, \delta_{\sigma\sigma'}\, \delta(\bar{x} - \bar{x}').$$

On the other hand, according to (21.13):

$$\left[\bar{\psi}_\sigma(\bar{x}, \bar{t}), \bar{\psi}_{\sigma'}(\bar{x}', \bar{t})\right]_+ = \left[\bar{\pi}_\sigma(\bar{x}, \bar{t}), \bar{\pi}_{\sigma'}(\bar{x}', \bar{t})\right]_+ = 0.$$

This proves the invariance of the commutations rules for simultaneous world points and for infinitesimal Lorentz transformations. The invariance with respect to any (non-infinitesimal) transformation follows directly from the group character of the Lorentz transformations, so that all invariance conditions are satisfied since evidently the Lagrangian (21.2) is also an invariant.

Nothing is changed in the quantization of the electron waves, if the electromagnetic field is considered as a variable wave field instead of a given field, as before. Φ_ν then stands for the field operators, which were denoted by ψ_ν in §17. One must add the Hamiltonian of the electromagnetic vacuum field to the Hamiltonian (21.9) which represents the kinetic energy of the electrons and their interaction with the electromagnetic field [corresponding to $\sum_n H_{(n)}$ in §17, cf. (17.9, 17.11)]. Furthermore, the Schrödinger function of the total system must satisfy certain subsidiary conditions [cf. (17.21)] which guarantee the validity of the Maxwell equations and which can be used to eliminate the longitudinal light waves, as was explained in §17. We shall not repeat this calculation, since it is essentially the same as that in §17 and since it leads exactly to the same result [cf. (17.46, 17.47)]: in the Hamiltonian only the transverse field components remain ($\Phi_0 \to 0$, $\Phi \to \Phi^{tr}$) and in the place of the energy of the longitudinal light waves appears the Coulomb energy of the electrons [cf. (17.35 to 17.39)]:

$$H^C = \frac{1}{8\pi}\int dx \int dx' \frac{\varrho(x)\,\varrho(x')}{|x - x'|},$$

where $\rho = -i\epsilon\pi\psi$.[1] Furthermore, the multiple-time formalism of quantum electrodynamics (§18) can also be carried through, the quantized electron wave field now taking the part of the individual electrons which were previously treated in configuration space. The time coordinate t_1 of the electron field replaces the particle times t_n, and it is different from the time t of the Maxwell field.[2] In order to avoid unnecessary complications, we shall deal in the following mainly with the case of the given field Φ_ν, but shall occasionally also consider the Maxwell field as a quantized field.

For the perturbation treatment of weak field effects, we shall write the Hamiltonian (21.9):

$$(21.16)\qquad \begin{cases} \mathbf{H} = \mathbf{H}^0 + \mathbf{H}', \\ \mathbf{H}^0 = -c\pi\{(\alpha\cdot\nabla) - i\mu\beta\}\psi, \\ \mathbf{H}' = i\varepsilon\pi\{(\alpha\cdot\Phi) - \Phi_0\}\psi. \end{cases}$$

If we expand, furthermore, ψ and π with respect to the field-free eigen functions u_m according to (21.11), it follows that:

$$H = \int dx\,\mathbf{H} = H^0 + H',$$

$$H^0 = \sum_m E_m a_m^* a_m,$$

$$(21.17)\qquad H' = \sum_{m,n} E'_{mn} a_m^* a_n, \quad \text{where } E'_{mn} = -\varepsilon h \int dx\, u_m^* \{(\alpha\cdot\Phi) - \Phi_0\} u_n.$$

Here it is permitted to rewrite H^0 in the sense of the Dirac hole theory by subtracting the constant $\sum_{m\,(E_m<0)} E_m$ [cf. (20.42, 20.43)]:

$$(21.18)\qquad H^0 = \sum_{m\,(E_m>0)} E_m N_m + \sum_{m\,(E_m<0)} |E_m| N'_m.$$

[1] It is to be noted that according to (21.10) the $\rho(x)$ and $\rho(x')$ commute: $[\rho(x), \rho(x')] = 0$; consequently the ρ_k and the $\rho_{k'}$, too, commute [cf. (17.27 ff.)].

[2] Cf. also S. Tomonaga, *Progr. Theor. Phys.* *1*, 40 (1946), and Z. Koba, T. Tati, and S. Tomonaga, *Progr. Theor. Phys.* *2*, 101 (1947), where the multiple-time formalism of quantum electrodynamics is still further generalized.

The perturbation function H' describes transitions, produced by the field Φ_ν, between the unperturbed (field-free) states; the term m, n in (21.17) corresponds here according to (20.18) to the transition of an electron from the single state n into the (initially not occupied) single state m. We shall interpret here the unoccupied levels in the negative energy spectrum again as positrons. A term m, n with $E_m > 0$ and $E_n < 0$ describes therefore the creation of an electron-positron pair: $N_m = 0 \to 1$, $N_n = 1 \to 0$ or $N'_n = 0 \to 1$ [cf. (20.43)]. We shall return later to a more precise and more general formulation of the positron theory, and shall instead first apply the perturbation method to the scattering of light by a free electron as a typical example.

For the matrix element of a two-step transition $\mathrm{I} \to \mathrm{II} \to \mathrm{F}$ we write as in (7.10):

$$H''_{\mathrm{F\,I}} = \sum_{\mathrm{II}} \frac{H'_{\mathrm{F\,II}} H'_{\mathrm{II\,I}}}{H^0_{\mathrm{I}} - H^0_{\mathrm{II}}} \tag{21.19}$$

We shall first consider a transition in which an initially existing light quantum of energy $h\omega_{\mathrm{I}}$ will be first absorbed by an electron in the single state m_{I} ($E_{m_{\mathrm{I}}} > 0$) and then emitted with the energy $h\omega_{\mathrm{F}}$. Such a transition contributes to H''_{FI}:[1]

$$\frac{H'_{\mathrm{F\,II}} H'_{\mathrm{II\,I}}}{H^0_{\mathrm{I}} - H^0_{\mathrm{II}}} = \frac{E'^{(\mathrm{Em})}_{m_{\mathrm{F}} m_{\mathrm{II}}} a^*_{m_{\mathrm{F}}} a_{m_{\mathrm{II}}} E'^{(\mathrm{Abs})}_{m_{\mathrm{II}} m_{\mathrm{I}}} a^*_{m_{\mathrm{II}}} a_{m_{\mathrm{I}}}}{(E_{m_{\mathrm{I}}} + h\,\omega_{\mathrm{I}}) - E_{m_{\mathrm{II}}}}$$
$$= \frac{E'^{(\mathrm{Em})}_{m_{\mathrm{F}} m_{\mathrm{II}}} E'^{(\mathrm{Abs})}_{m_{\mathrm{II}} m_{\mathrm{I}}}}{(E_{m_{\mathrm{I}}} + h\,\omega_{\mathrm{I}}) - E_{m_{\mathrm{II}}}} a^*_{m_{\mathrm{F}}} a_{m_{\mathrm{I}}} \left(1 - N_{m_{\mathrm{II}}}\right) \tag{21.20}$$

[cf. (20.18) ff.]. Single states m_{II} which are occupied in the beginning ($N_{m_{\mathrm{II}}} = 1$) contribute nothing to (21.20); the respective transitions $\mathrm{I} \to \mathrm{F}$ are not permitted by the Pauli principle. This is true in particular for all negative single states ($E_{m_{\mathrm{II}}} < 0$), if one assumes that there exists in the beginning no positron, i.e., no hole in the negative spectrum. Instead, the initially occupied levels give rise to transitions of the following kind: first the light quantum $h\,\omega_{\mathrm{F}}$ is emitted, whereby the initially existing electron m_{II} jumps onto the final level m_{F}.

[1] It is convenient not to insert special numerical values for the matrix elements of a_n, a^*_m but to consider H''_{FI} still as a matrix with respect to the occupation numbers $N_1, N_2 \ldots$ [just as (9.19) was considered as a matrix with respect to the charge numbers $\lambda_1, \lambda_2 \ldots$].

After this the initial electron m_{I} falls into the produced hole m_{II} with absorption of the initial light quantum $h\,\omega_{\mathrm{I}}$:

$$(21.21) \qquad \frac{H'_{\mathrm{F\,II}}\,H'_{\mathrm{II\,I}}}{H^0_{\mathrm{I}} - H^0_{\mathrm{II}}} = \frac{E'^{(\mathrm{Abs})}_{m_{\mathrm{II}}\,m_{\mathrm{I}}}\,\overset{*}{a}_{m_{\mathrm{II}}}\,a_{m_{\mathrm{I}}}\,E'^{(\mathrm{Em})}_{m_{\mathrm{F}}\,m_{\mathrm{II}}}\,\overset{*}{a}_{m_{\mathrm{F}}}\,a_{m_{\mathrm{II}}}}{E_{m_{\mathrm{II}}} - (E_{m_{\mathrm{F}}} + h\omega_{\mathrm{F}})}$$

$$= \frac{E'^{(\mathrm{Em})}_{m_{\mathrm{F}}\,m_{\mathrm{II}}}\,E'^{(\mathrm{Abs})}_{m_{\mathrm{II}}\,m_{\mathrm{I}}}}{(E_{m_{\mathrm{F}}} + h\omega_{\mathrm{F}}) - E_{m_{\mathrm{II}}}}\,\overset{*}{a}_{m_{\mathrm{F}}}\,a_{m_{\mathrm{I}}}N_{m_{\mathrm{II}}}$$

[The sign results from the fact that the matrices a and a^* are anticommutative. The matrix elements $E'^{(\mathrm{Em})}_{m_{\mathrm{F}}\,m_{\mathrm{II}}}$ and $E'^{(\mathrm{Abs})}_{m_{\mathrm{II}}\,m_{\mathrm{I}}}$ have here the same meaning as in (21.20)]. Since the scattering process takes place with energy conservation, it follows that:

$$E_{m_{\mathrm{F}}} + h\,\omega_{\mathrm{F}} = E_{m_{\mathrm{I}}} + h\,\omega_{\mathrm{I}},$$

and consequently the sum of the two terms (21.20 and 21.21) is independent of $N_{m_{\mathrm{II}}}$, namely, equal to:

$$\frac{E'^{(\mathrm{Em})}_{m_{\mathrm{F}}\,m_{\mathrm{II}}}\,E'^{(\mathrm{Abs})}_{m_{\mathrm{II}}\,m_{\mathrm{I}}}}{(E_{m_{\mathrm{I}}} + h\,\omega_{\mathrm{I}}) - E_{m_{\mathrm{II}}}}\,\overset{*}{a}_{m_{\mathrm{F}}}\,a_{m_{\mathrm{I}}};$$

i.e., each level m_{II} contributes, whether occupied or unoccupied, the same amount to the sum (21.19). The same holds, as can easily be seen, for another class of transitions, which differ from the ones discussed so far in the sequence of the light emission and light absorption processes, other things being equal. Altogether, one obtains:

$$(21.22) \qquad H''_{\mathrm{FI}} = \sum_{m_{\mathrm{II}}}\left\{\frac{E'^{(\mathrm{Em})}_{m_{\mathrm{F}}\,m_{\mathrm{II}}}\,E'^{(\mathrm{Abs})}_{m_{\mathrm{II}}\,m_{\mathrm{I}}}}{(E_{m_{\mathrm{I}}} + h\,\omega_{\mathrm{I}}) - E_{m_{\mathrm{II}}}} + \frac{E'^{(\mathrm{Abs})}_{m_{\mathrm{F}}\,m_{\mathrm{II}}}\,E'^{(\mathrm{Em})}_{m_{\mathrm{II}}\,m_{\mathrm{I}}}}{E_{m_{\mathrm{I}}} - (E_{m_{\mathrm{II}}} + h\,\omega_{\mathrm{F}})}\right\}\overset{*}{a}_{m_{\mathrm{F}}}\,a_{m_{\mathrm{I}}}.$$

Insofar as the scattering process is not altogether forbidden by the exclusion principle (i.e., provided that for the initial state $N_{m_{\mathrm{I}}} = 1$, $N_{m_{\mathrm{F}}} = 0$) its probability is, according to (21.22), the same as that calculated with the unquantized Dirac wave mechanics (without hole theory); in other words, it will be given by the well-known formula of Klein and Nishina.[1] In the unrelativistic limiting case, it will reduce to the classical (Thomson) scattering formula (cf. §17). The same is also true for the scattering of light on positrons.

Further problems which could be treated with the perturbation theory, on the basis of (21.17), are the creation of an electron-positron pair by light and the reverse process, the annihilation of a pair. On account of the conservation of energy and momentum, these processes—as was explained in §11 for the creation of a meson pair—depend on the presence of an additional electromagnetic field (for instance, an electrostatic field).[2]

The perturbation method, as outlined above, yields satisfactory results only for a very limited class of problems; in other cases—even irrespective of the quantum electrodynamical self-energies—the sums over the virtual, intermediate states diverge. An example is the vacuum polarization which arises, as in the case of the scalar charged field (cf. §11), from the fact that the switching on of an electric field in "vacuum" produces virtual electron transitions (pair creation); the charge-density produced by this perturbation is infinite, even after subtraction of the field-free "vacuum" charge. This raises the question how to arrive at reasonable definitions for the charge- and energy-momentum quantities of the electrons in electric fields. We owe to Dirac[3] and Heisenberg[4] a formulation for a subtraction rule, which seems satisfactory and which, in addition, is symmetrical with regard to the sign of the electrical charge (cf. §20). We can give here only a short review of this method.

We introduce a density matrix $r_{\sigma\sigma'}(x, t; x', t')$, just as in the field-free case, which shall depend on the field operators ψ_σ, π_σ, as well as on the electromagnetic field functions in a way which shall be determined later. With the help of this matrix, current- and energy-momentum quantities are to be represented as follows:

$$s_\nu = \varepsilon h c \sum_{\sigma, \sigma'} \gamma^{(\nu)}_{\sigma'\sigma} r_{\sigma\sigma'}, \tag{21.23}$$

$$T_{\mu\nu} = \frac{1}{2} (\Theta_{\mu\nu} + \Theta_{\nu\mu}), \qquad \Theta_{\mu\nu} = \frac{hc}{2i} (\partial_\nu - \partial_\nu'^{*}) \sum_{\sigma, \sigma'} \gamma^{(\mu)}_{\sigma'\sigma} r_{\sigma\sigma'}, \tag{21.24}$$

---

[1] *Z. Phys.* *52*, 853, 1929. Cf. also Heitler, *Quantum Theory of Radiation*, Oxford Univ. Press, London, 1936, §16.

[2] Oppenheimer and Plesset, *Phys. Rev.* *44*, 53, 1933; Heitler and Sauter, *Nature* *132*, 892, 1933; Bethe and Heitler, *Proc. Roy. Soc. London* *146*, 83, 1934. Cf. also Heitler, loc. cit., §20. The most important process for pair annihilation is that with the emission of two light quanta (Dirac, *Proc. Cambridge Phil. Soc.* *26*, 361, 1930), cf. Heitler, loc. cit. §21.

[3] *Proc. Cambridge Phil. Soc.* *30*, 150, 1934.

[4] *Z. Phys.* *90*, 209, 1934.

where ∂_ν, $\partial_\nu^{\prime *}$ according to (21.1) have the following significance:

$$(21.25) \qquad \partial_\nu = \frac{\partial}{\partial x_\nu} - \frac{i\varepsilon}{c}\Phi_\nu(x, t), \qquad \partial_\nu^{\prime *} = \frac{\partial}{\partial x_\nu^{\prime}} + \frac{i\varepsilon}{c}\Phi_\nu(x', t').$$

These definitions are for $\Phi_\nu = 0$ slightly more general than the definitions (20.48, 20.49) used in §20 in that we do not yet require $t' \rightarrow t$, $x' \rightarrow x$. With (21.23, .24) the s_ν and $T_{\mu\nu}$ are now defined as matrices with respect to space and time coordinates. The proper physical density functions shall be defined later by a suitable limiting process, as finite limiting values. In order to satisfy the continuity equation for the 4-current, it is sufficient to require that the density matrix r should satisfy the differential equation:

$$(21.26) \qquad \sum_\nu (\partial_\nu + \partial_\nu^{\prime *}) \sum_{\sigma, \sigma'} \gamma_{\sigma'\sigma}^{(\nu)} r_{\sigma\sigma'} = 0;$$

for then $\sum_\nu (\partial_\nu + \partial_\nu^{\prime *}) s_\nu = 0$, or in the limit $x', t' \rightarrow x, t$ according to (21.25):

$$\sum_\nu \frac{\partial}{\partial x_\nu} (\lim s_\nu) = 0.$$

It follows further from (21.26):

$$\sum_\mu (\partial_\mu + \partial_\mu^{\prime *})\, \Theta_{\mu\nu} = \frac{hc}{2i} \sum_\mu \left[(\partial_\mu + \partial_\mu^{\prime *}), (\partial_\nu - \partial_\nu^{\prime *})\right] \sum_{\sigma, \sigma'} \gamma_{\sigma'\sigma}^{(\mu)} r_{\sigma\sigma'};$$

considering that according to (21.25) [cf. (13.6)]:

$$\begin{aligned}\left[(\partial_\mu + \partial_\mu^{\prime *}), (\partial_\nu - \partial_\nu^{\prime *})\right] &= \left[\partial_\mu, \partial_\nu\right] - \left[\partial_\mu^{\prime *}, \partial_\nu^{\prime *}\right] \\ &= -\frac{i\varepsilon}{c}\left\{F_{\mu\nu}(x, t) + F_{\mu\nu}(x', t')\right\}\end{aligned}$$

it follows:

$$\sum_\mu (\partial_\mu + \partial_\mu^{\prime *})\, \Theta_{\mu\nu} = -\frac{1}{2c} \sum_\mu \left\{F_{\mu\nu}(x, t) + F_{\mu\nu}(x', t')\right\} s_\mu,$$

and in the limit $x', t' \rightarrow x, t$:

$$\sum_\mu \frac{\partial}{\partial x_\mu} (\lim \Theta_{\mu\nu}) = -\frac{1}{c} \sum_\mu s_\mu F_{\mu\nu}.$$

In order that also:

$$\sum_{\mu} \frac{\partial}{\partial x_\mu} (\lim T_{\mu\nu}) = -\frac{1}{c} \sum_{\mu} s_\mu F_{\mu\nu}$$

be true, one must assume besides (21.26) that:

$$(21.27) \quad \sum_{\mu} \frac{\partial}{\partial x_\mu} \{\lim (\Theta_{\mu\nu} - \Theta_{\nu\mu})\} = \lim \sum_{\mu} (\partial_\mu + \partial_\mu'^*) (\Theta_{\mu\nu} - \Theta_{\nu\mu}) = 0.$$

If we chose in (21.23, 21.24):

$$r_{\sigma\sigma'} (x, t; x', t') = \psi^\dagger_{\sigma'} (x', t') \psi_\sigma (x, t)$$

this would lead us back to the equations (21.4, 21.5) in which no subtraction of vacuum terms has yet taken place. In order to correct the density matrix in accordance with the general ideas of the hole theory, we follow Dirac and Heisenberg in assuming, as in the field-free case [cf. (20.51, 20.54)]:

$$(21.28) \quad r = R + S,$$

$$(21.29) \quad R_{\sigma\sigma'} (x, t; x', t') = \frac{1}{2} \{\psi^\dagger_{\sigma'} (x', t') \psi_\sigma (x, t) - \psi_\sigma (x, t) \psi^\dagger_{\sigma'} (x', t')\}$$

($\psi^\dagger = i \psi^* \beta = \pi \beta/h$). According to the Dirac equation (21.3) the matrix R satisfies the differential equations:

$$\sum_{\nu} \sum_{\varrho} \gamma^{(\nu)}_{\sigma\varrho} \partial_\nu R_{\varrho\sigma'} + \mu R_{\sigma\sigma'} = 0,$$

$$\sum_{\nu} \sum_{\varrho} \partial_\nu'^* R_{\sigma\varrho} \gamma^{(\nu)}_{\varrho\sigma'} - \mu R_{\sigma\sigma'} = 0.$$

By addition of these equations and contraction with regard to the σ, σ' it follows further that:

$$\sum_{\nu} (\partial_\nu + \partial_\nu'^*) \sum_{\varrho, \sigma} \gamma^{(\nu)}_{\varrho\sigma} R_{\sigma\varrho} = 0,$$

i.e., the R-part of r satisfies the equation (21.16). The R-part also agrees with the postulate (21.27), as can easily be verified.[1] The S-part, which is given by (20.55, 20.56) in the field-free case, must satisfy the same conditions. One must further require that S should be independent of the state of the electrons; i.e., S must not contain the operators ψ, $\psi^\dagger$, while it may explicitly

[1] Cf. footnote page 189.

depend on the electromagnetic potential functions and their derivatives, subject, of course, to the condition of relativistic and gauge invariance.

Our task now is to determine the matrix S subject to the above-mentioned restrictions such that the operator $r = R + S$ will yield finite results in the limit $x', t' \to x, t$. Only those terms of S are important which influence the limiting values of s_ν and $T_{\mu\nu}$ $(x', t' \to x, t)$. In this sense Heisenberg[1] was essentially able to solve the problem unambiguously, after Dirac had already prepared the way. We cannot enter into the rather lengthy calculations and shall only describe shortly the corrected Hamiltonian $H = \int dx\,(-T_{44} + \rho\,\Phi_0)$ as derived by Heisenberg. It contains besides the terms (21.16), subtraction terms which are unit matrices with regard to the electron numbers N_m and which depend on the Φ_ν and their derivatives in a complicated way. Φ_ν here appears always multiplied by the elementary charge $\epsilon\,h$. If H is expanded in powers of the electron charge:

$$H = \sum_k H^{(k)},$$

then $H^{(0)}$ agrees, of course, with the field-free values (20.42, 20.43), or (21.18). If the Maxwell field is quantized then the light quantum energy must be added to this. We obtain, moreover:

$$H^{(1)} = \sum_{\substack{m \\ (E_m > 0)}} E'_{mm} N_m - \sum_{\substack{m \\ (E_m < 0)}} E'_{mm} N'_m + \sum_{m \neq n} E'_{mn} \overset{*}{a}_m a_n ,$$

where E' is the matrix defined in (21.17). As far as terms $m \neq n$ are concerned, $H^{(1)}$ agrees, of course, with the perturbation function H' (21.17), which was used earlier. This justifies the perturbation theory used above for light scattering, pair creation, etc., since the Heisenberg perturbation function of the second order $H^{(2)}$ gives no contribution to the respective matrix elements. The method, however, becomes much more complicated for problems which involve higher order approximations. While for $H^{(0)}$ and $H^{(1)}$ there was no difficulty in stating just their limiting values $x', t' = x, t$, a description of the higher order terms requires going back to the density functions $\mathsf{H}^{(2)}, \ldots$ which moreover must still be considered as matrices with respect to the space-time coordinates since their limiting values are not necessarily finite. In order to

[1] Heisenberg, loc. cit. There seems to exist a gap, in so far as it has not been proved that the equation (21.27) can be satisfied for the S-part.

obtain from these the integral quantities $H^{(k)}$, we put:

$$x = \bar{x} + \xi, \qquad x' = \bar{x} - \xi$$

—to simplify we choose $t' = t$—and integrate over the $\bar{x}$-space with constant vector ξ:

$$H^{(k)} = \int d\bar{x}\, \mathsf{H}^{(k)}\, (\bar{x} + \xi,\ \bar{x} - \xi).$$

The limiting process $\xi \rightarrow 0$ can in general only be carried out after the perturbation calculation, in a second approximation, for instance, in the matrix elements:

$$H^{(2)}_{\mathrm{F\,I}} + \sum_{\mathrm{II}} \frac{H^{(1)}_{\mathrm{F\,II}}\ H^{(1)}_{\mathrm{II\,I}}}{H^{(0)}_{\mathrm{I}} - H^{(0)}_{\mathrm{II}}};$$

in that case it leads to finite results. The R-terms do not contribute anything to $\mathsf{H}^{(2)}$, $\mathsf{H}^{(3)}$, . . . ; hence each of the higher perturbation functions ($k \geqq 2$) is constant (unit matrix) with respect to the electron and positron numbers N_m, N'_m, while their dependence on the potentials Φ_ν is determined by that of the S-matrix. As a (relatively simple) example we shall write down $\mathsf{H}^{(4)}$:

$$\mathsf{H}^{(4)} = \frac{1}{48\,\pi^2}\left(\frac{\varepsilon^2\, h}{c}\right)^2 \frac{1}{h\, c}\left(\Phi\,(x)\cdot\frac{\xi}{|\xi|}\right)^4.$$

$\mathsf{H}^{(5)}$ and the higher terms vanish for $\xi = 0$.

The Heisenberg terms $H^{(2)}$, $H^{(3)}$, $H^{(4)}$ are principally important for electrodynamic problems. Since the Hamiltonian contains terms of the third and fourth order in Φ_ν, the electromagnetic field equations are no longer strictly linear, i.e., one must expect deviations from the principle of superposition for high field strengths. Intuitively this may be interpreted as a reaction of the vacuum polarization on the field, which is of a non-linear character (as in a medium which can be polarized, and which has a dielectric constant depending on the field strength). As a typical non-linear effect we mention the scattering of light on light or on an electric field. Such effects can already be understood qualitatively from the primitive picture of the holes.[1] Thus, for instance, one interprets the interaction of two light quanta, which gives rise to their scattering, as due to the emission and reabsorption of virtual electron-positron pairs in a similar way as two electrons are supposed to interact by emission and reabsorption of light quanta. Heisenberg's subtraction formalism is needed, however,

[1] Halpern, *Phys. Rev.* 44, 855, 1933; Delbrück, *Z. Phys.* 84, 144, 1933.

to obtain a sufficiently small probability for the scattering of light on light (especially at low frequencies).[1] In the respective matrix element, which is of the fourth order in ϵ, the contributions from the perturbation function $H^{(1)}$ are chiefly compensated by the term $H^{(4)}$. (This term contains, evidently, matrix elements which correspond to the annihilation and creation of two light quanta without change in the state of the electrons. Φ in $H^{(4)}$ is, of course, to be considered as the field operator, called ψ in §16). Also for the scattering of light on electric fields (at not too high frequencies) only a very small intensity results which is hardly of measurable order of magnitude.[2] Provided that all wave lengths involved are large compared with the Compton wave length of the electron, it seems possible to describe all these effects which are not dependent on the state of the electrons by a new Lagrangian for the electromagnetic field:[3]

$$L = \frac{1}{2}(\mathfrak{E}^2 - \mathfrak{H}^2) + \frac{1}{360\,\pi^2}\frac{\varepsilon^4\, h^5}{m^4\, c^7}\{(\mathfrak{E}^2 - \mathfrak{H}^2)^2 + 7\,(\mathfrak{E}\cdot\mathfrak{H})^2\} + \cdots.$$

(This resembles the Lagrangian of the non-linear theories of Mie and Born, which aim at a "unitary" description of field and charged particle. Whether this connection is more than purely formal seems doubtful in view of the difference of the fundamental ideas in the two theories.)

The Dirac-Heisenberg subtraction formalism has, however, not led to any improvement regarding the self-energy problem of quantum electrodynamics. The electromagnetic self-energy of the electron is still infinite as before, although the respective momentum space integral now diverges only logarithmically.[4] But just this fact makes it appear doubtful whether a solution or a reduction of the problem can be achieved by applying the multiple-time theory, as in the configuration-space theory of the electrons (cf. §19). Furthermore, the light quantum, too, gives rise to a self-energy in the "hole" theory, on account of its ability to create virtual electron-positron pairs. Here again the Heisenberg formalism leads to a logarithmically divergent integral.[5] If the momentum spectra are cut off, the self-energy terms prove to be small in the second approximation, compared with the unperturbed energy eigen values (E_m or $h\,c|k|$), even if the cut-off momentum is chosen $\gg m\,c$. The reason is that,

---

[1] Euler, *Ann. Physik 26*, 398, 1936.

[2] Kemmer, *Helv. Phys. Acta 10*, 112, 1937.

[3] Euler, *loc. cit.;* Heisenberg and Euler, *Z. Phys. 98*, 714, 1936; Weisskopf, *Kgl. Danske Vidensk. Selsk., Math.-fys. Medd. XIV*, 6, 1936; Kemmer, *loc. cit.*

[4] Weisskopf, *Z. Phys. 89*, 27, 1934, and *90*, 817, 1934; *Phys. Rev. 56*, 72, 1939.

[5] Heisenberg, *loc. cit.*

on the one hand, this momentum enters only logarithmically, and that, on the other hand, the pure number $\epsilon^2 h/c$ which enters as an expansion parameter is small. As was said before, this method of cutting off is only a makeshift theory because it destroys the Lorentz invariance of the theory.

Chapter VI

Supplementary Remarks

§ 22. Particles with Higher Spin. Spin and Statistics

The types of field which were discussed extensively in the preceding chapters were selected partly on account of their relative simplicity, partly on account of their connections, either assumed or verified, with elementary particles known from experiments: mesons, photons, and electrons. Regarding other, more complicated types of field, we must be content with a short report, especially in view of the rapidly growing complications of the mathematical formalism for fields with more components. So far there is no indication that such higher fields are realized in nature, with the only exception of "gravitational waves," about which we shall speak later.

As we have seen, there exists a connection between the relativistic transformation properties of the field and the spin of the particles which are described by the quantized field: the scalar field describes particles with spin 0, the vector fields those with spin 1, while electrons with spin $\frac{1}{2}$ are represented by a Dirac "spinor" field. The generalization for integral spin is obvious: we use tensors of rank s to describe particles with spin s. The van der Waerden "spinor calculus,"[1] which will not be discussed here, provides for half-integral spins ($s = 3/2, 5/2, \ldots$) the formal mathematical apparatus for the construction of field functions with the desired transformation properties. This calculus allows even a uniform description of particles with integral and non-integral spin.[2] But in the case of integral spin the respective spinors can be reduced to tensors.

The tensor or spinor fields must satisfy certain additional conditions if they shall describe only particles of a definite spin s. We have already seen an

[1] Van der Waerden, *Die gruppentheoretische Methode in der Quantenmechanik*, Springer, Berlin, 1932.

[2] Dirac, *Proc. Roy. Soc. London*, *155*, 447, 1936; Fierz, *Helv. Phys. Acta 12*, 3, 1939.

example for this in §12. We had to subject the vector field ψ_ν to the subsidiary condition (12.2) (vanishing of the 4-divergence) in order to avoid having particles with spin 0 appear besides those with spin 1. Since the former would have in addition a negative energy, the condition (12.2) simultaneously provides that the field energy is positive-definite. The additional conditions serve the same purpose for cases of higher spin.

We shall now discuss briefly the case $s = 2$ in order to indicate the generalization of the formalism for higher spin values.[1] Let $\psi_{\mu\nu}$ be a symmetrical tensor of the second rank with identically vanishing trace:

$$\psi_{\mu\nu} = \psi_{\nu\mu}, \qquad \sum_\nu \psi_{\nu\nu} = 0. \tag{22.1}$$

The Schrödinger-Gordon equation shall hold for each field component $\psi_{\mu\nu}$:

$$(\Box - \mu^2)\,\psi_{\mu\nu} = 0, \tag{22.2}$$

In addition, one requires, similar to (12.2), that the vector divergence of the tensor ψ vanishes:

$$\sum_\mu \frac{\partial \psi_{\mu\nu}}{\partial x_\mu} = 0. \tag{22.3}$$

We shall write ψ—in the non-quantized theory—as a plane wave:

$$\psi_{\mu\nu} = a_{\mu\nu}\, e^{ikx - i\omega_k t}, \quad \text{where} \quad \omega_k^2 = c^2\,(\mu^2 + k^2),$$

and we shall count how many independent waves exist for a given wave number vector k. If we use, for reasons of simplicity, the reference system in which $k = 0$ (rest-system of the respective particles in the quantized theory), there results:

$$\psi_{\mu\nu} = a_{\mu\nu}\, e^{-i\mu t},$$

and the divergence condition (22.3) yields:

$$\mu \cdot \psi_{4\nu} = 0.$$

If the rest mass is assumed $\mu \neq 0$, then in this coordinate system all components $\psi_{4\nu} = \psi_{\nu 4}$ will vanish, and there remain only the independent components of the three dimensional tensor ψ_{ik} with the properties:

$$\psi_{ik} = \psi_{ki}, \qquad \sum_k \psi_{kk} = 0$$

[1] We follow here the representation of Fierz, *loc. cit.*

The number of the linearly-independent components is obviously 5, i.e., equals $2s + 1$. It can easily be deduced from the way in which these components transform under a rotation of the space coordinate system that the appertaining five particle states in the quantized theory correspond exactly to the five possible orientations of the spin 2.

The case of vanishing rest mass, $\mu = 0$, must be treated separately, just as for $s = 1$ (electromagnetic field, cf. §16). Then there exist "gauge transformations" [analogous to (16.4)] which leave the field equations invariant. The equations (22.1, 22.2, 22.3), for instance, with $\mu = 0$, remain unchanged, if $\psi_{\mu\nu}$ is replaced by $\psi_{\mu\nu} + \frac{\partial \Lambda_\nu}{\partial x_\mu} + \frac{\partial \Lambda_\mu}{\partial x_\nu}$, where Λ_ν is a vector field which satisfies the conditions $\Box \Lambda_\nu = 0$ and $\sum_\nu \partial \Lambda_\nu / \partial x_\nu = 0$. If one now considers two states of the field as physically equivalent which transform into each other by a gauge transformation, then there remain only two independent, non-equivalent "states of polarization," for a definite frequency and direction of a plane wave. This is true for all spin values $s \neq 0$.

The problem of constructing Lagrangians in such a way that the wave equation and the subsidiary conditions follow simultaneously from the Euler differential equations is quite difficult.[1] The simplest assumption for $s = 2$ is:

$$L = -\sum_{\lambda,\mu,\nu} \frac{\partial \psi^*_{\mu\nu}}{\partial x_\lambda}\left(\frac{\partial \psi_{\mu\nu}}{\partial x_\lambda} - 2\frac{\partial \psi_{\lambda\nu}}{\partial x_\mu}\right) - \sum_{\mu,\nu}\left(\frac{\partial \psi^*_{\mu\nu}}{\partial x_\mu}\frac{\partial \psi}{\partial x_\nu} + \frac{\partial \psi^*}{\partial x_\nu}\frac{\partial \psi_{\mu\nu}}{\partial x_\mu}\right) + \sum_\nu \frac{\partial \psi^*}{\partial x_\nu}\frac{\partial \psi}{\partial x_\nu} - \mu^2\left(\sum_{\mu,\nu} \psi^*_{\mu\nu}\psi_{\mu\nu} - \psi^*\psi\right); \tag{22.4}$$

here $\psi_{\mu\nu}$ is meant to be a symmetrical tensor right from the beginning and ψ has the significance:

$$\psi = \sum_\nu \psi_{\nu\nu}. \tag{22.5}$$

The variation with respect to the ten independent field components $\psi^*_{\mu\nu}$ yields the field equations:

[1] Cf. Fierz and Pauli, *Proc. Roy. Soc.* *173*, 211, 1939. The Lagrangian functions mentioned in this paper also contain auxiliary fields which belong to smaller spin and whose disappearance also can be derived subsequently from the principle of variation. The influence of an external electromagnetic field can be taken into account by replacing in these Lagrangian functions $\partial/\partial x_\nu$ by ∂_ν [cf. (13.2), (21.1)]. We disregard here external forces, i.e., we discuss only "vacuum fields."

$$(22.6)\quad \begin{cases} \Box\,\psi_{\mu\nu} - \sum\limits_{\lambda}\left(\dfrac{\partial^2 \psi_{\lambda\nu}}{\partial x_\lambda\,\partial x_\mu} + \dfrac{\partial^2 \psi_{\lambda\mu}}{\partial x_\lambda\,\partial x_\nu}\right) + \dfrac{\partial^2 \psi}{\partial x_\mu\,\partial x_\nu} \\ + \delta_{\mu\nu}\left(\sum\limits_{\varkappa,\lambda} \dfrac{\partial^2 \psi_{\varkappa\lambda}}{\partial x_\varkappa\,\partial x_\lambda} - \Box\,\psi\right) - \mu^2\left(\psi_{\mu\nu} - \delta_{\mu\nu}\,\psi\right) = 0. \end{cases}$$

If one applies here on the left side successively the operators:

$$\sum_{\mu} \frac{\partial}{\partial x_\mu}\cdots,\quad \sum_{\mu,\nu} \frac{\partial^2}{\partial x_\mu\,\partial x_\nu}\cdots,\quad \sum_{\mu,\nu} \delta_{\mu\nu}\cdots.$$

there follows:

$$(22.7)\qquad \mu^2\left(\sum_{\mu} \frac{\partial \psi_{\mu\nu}}{\partial x_\mu} - \frac{\partial \psi}{\partial x_\nu}\right) = 0,$$

$$(22.8)\qquad \mu^2\left(\sum_{\mu,\nu} \frac{\partial^2 \psi_{\mu\nu}}{\partial x_\mu\,\partial x_\nu} - \Box\,\psi\right) = 0,$$

$$(22.9)\qquad 2\sum_{\mu,\nu} \frac{\partial^2 \psi_{\mu\nu}}{\partial x_\mu\,\partial x_\nu} - (2\,\Box - 3\,\mu^2)\,\psi = 0.$$

For $\mu \neq 0$ it follows from (22.8 and 22.9):

$$(22.10)\qquad \psi = 0,$$

and hence from (22.7):

$$(22.11)\qquad \sum_{\mu} \frac{\partial \psi_{\mu\nu}}{\partial x_\mu} = 0.$$

In view of this (22.6) implies:

$$(22.12)\qquad (\Box - \mu^2)\,\psi_{\mu\nu} = 0.$$

The equations (22.5, 22.10, 22.11, 22.12) agree with (22.1, 22.2, 22.3) as was desired.

For $\mu = 0$, on the other hand, the equations (22.10, 22.11, 22.12) no longer follow from (22.6), since the equations (22.7, 22.8) degenerate to identities. This corresponds to the earlier result for $s = 1$. By putting $\mu = 0$, the equations (22.6, 22.9) go over into the Einstein differential equations for a weak gravitational field in a space free of matter; $\psi_{\mu\nu}$ stands for the deviation of the metrical fundamental tensor $g_{\mu\nu}$ from the unity tensor:

$$g_{\mu\nu} = \delta_{\mu\nu} + \psi_{\mu\nu},$$

and the terms quadratical in $\psi_{\mu\nu}$ are neglected.[1] If one introduces further the field:

$$\psi'_{\mu\nu} = \psi_{\mu\nu} - \frac{1}{2}\delta_{\mu\nu}\psi$$

one can, as Hilbert has shown,[1] transform the coordinates, without altering the infinitesimal character of $\psi_{\mu\nu}$, in such a way that:

$$\sum_\mu \frac{\partial \psi'_{\mu\nu}}{\partial x_\mu} = 0$$

This corresponds to a gauge transformation of the field $\psi_{\mu\nu}$.[2] The equations (22.6, 22.9) now reduce to:

$$\Box \psi'_{\mu\nu} = 0.$$

The plane waves which satisfy these equations have only two independent, non-equivalent states of polarization[3]; these "gravitational waves" are identical with the above mentioned solutions of the Fierz field equations for $s = 2$, $\mu = 0$. The corresponding particles of the quantized theory, the "gravitational

[1] Cf., for instance, Pauli, *Encykl. d. Math. Wissensch.* Teubner, Leipzig-Berlin, 1921, Vol. V, part 2, section 60. $\psi_{\mu\nu}$ is here, of course, a "real" tensor.

[2] The equations (22.6, 22.9) (with $\mu = 0$) are invariant under the gauge transformations:

$$\psi_{\mu\nu} \rightarrow \psi_{\mu\nu} + \frac{\partial \Lambda_\nu}{\partial x_\mu} + \frac{\partial \Lambda_\mu}{\partial x_\nu},$$

where Λ_ν may be any vector field. The field ψ' transforms according to:

$$\psi'_{\mu\nu} \rightarrow \psi'_{\mu\nu} + \frac{\partial \Lambda_\nu}{\partial x_\mu} + \frac{\partial \Lambda_\mu}{\partial x_\nu} - \delta_{\mu\nu} \sum_\varrho \frac{\partial \Lambda_\varrho}{\partial x_\varrho},$$

so that:

$$\sum_\mu \frac{\partial \psi'_{\mu\nu}}{\partial x_\mu} \rightarrow \sum_\mu \frac{\partial \psi'_{\mu\nu}}{\partial x_\mu} + \Box \Lambda_\nu;$$

by a suitable choice of Λ_ν this divergence and also simultaneously the trace $\psi' = \sum_\nu \psi'_{\nu\nu}$ can be made to vanish.

[3] Einstein, *Berliner Berichte 1918*, 154.

quanta," thus have a spin of the magnitude $2h$, with two allowed orientations (parallel or antiparallel to the direction of propagation).

A further task is to construct for any spin value energy-momentum tensors which satisfy the conservation equations (2.7); and, further, to define charge- and current-density functions for complex (charged) fields in accordance with the continuity equations (3.12). This has been carried out first by Jauch and Fierz.[1] Especially important are the results concerning the sign of energy and charge. We have seen earlier that the energy density is positive-definite for the cases $s = 0$ and $s = 1$. This is no longer true for higher spin values; although the total energy is for integral spin always positive-definite, while the total charge is indefinite. On the other hand, for non-integral spins, and in particular for $s = \frac{1}{2}$ (cf. §20), the charge is definite, but the energy indefinite: the sign of the total energy enters symmetrically into the theory. This has the same consequences for the quantization of these fields, as we met already in the theory of the electrons and positrons (see below).

[1] Fierz, *loc. cit.* In the force-free case, there are (except for the spin $s = 0$) several possible definitions for $T_{\mu\nu}$ and s_ν, for which the integrals $\int dx\, T_{4\nu}$ and $\int dx\, s_4$ (energy, momentum, and charge) are the same. The localization of these quantities in the field seems ambiguous. For instance, one can add, for $s = 1$, to the current density (12.7) a "polarization-current":

$$s'_\nu = \gamma \cdot i\, \varepsilon \sum_\mu \frac{\partial}{\partial x_\mu} \left(\overset{*}{\psi}_\nu \psi_\mu - \overset{*}{\psi}_\mu \psi_\nu \right)$$

(γ = const.) without violating the continuity equation. This corresponds to a modification of the Lagrangian by a divergence:

$$L \to L - \gamma \sum_\nu \frac{\partial}{\partial x_\nu} \sum_\mu \left(\frac{\partial \overset{*}{\psi}_\nu}{\partial x_\mu} \psi_\mu - \frac{\partial \overset{*}{\psi}_\mu}{\partial x_\mu} \psi_\nu \right)$$

$$= L - \gamma \sum_{\nu, \mu} \left(\frac{\partial \overset{*}{\psi}_\nu}{\partial x_\mu} \frac{\partial \psi_\mu}{\partial x_\nu} - \frac{\partial \overset{*}{\psi}_\mu}{\partial x_\mu} \frac{\partial \psi_\nu}{\partial x_\nu} \right).$$

For $s = \frac{1}{2}$ the corresponding substitution is:

$$L \to L - \text{const.} \sum_\nu \frac{\partial}{\partial x_\nu} \sum_\mu \psi^\dagger\, i \left(\gamma^{(\mu)} \gamma^{(\nu)} - \gamma^{(\nu)} \gamma^{(\mu)} \right) \frac{\partial \psi}{\partial x_\mu}.$$

The density definitions become unambiguous if one discusses the (charged) particles in the electromagnetic field and if one assumes their equations of motion in the field in a certain way. For $s = 1$, it means disposing of the constant γ in the additional term to the Lagrangian, mentioned in §13 (page 95), i.e., choosing the magnetic spin moment of the particle. A corresponding situation results for $s = \frac{1}{2}$.

The spin values $s > 1$ give rise to additional complications in the quantized theory. In the scalar theory (§§6 and 8) the canonical commutation rules (1.7) or (3.9) yielded directly a relativistic, invariant prescription for quantization. For $s = 1$ (§12) the subsidiary conditions (12.2) had the consequence that the canonical commutation rules could not be applied to the redundant field component ψ_4. Yet ψ_4 could be eliminated and thereupon the canonical formalism would yield again invariant commutation rules. For higher spin values the number of subsidiary conditions and with them also the number of redundant field components increases quickly, and their elimination is no longer possible. But a relativistic invariant quantization is feasible even without the canonical formalism. Fierz (loc. cit.) has found invariant commutation rules for any spin value which for $s \leqq 1$ are identical with the above derived relations (6.8) or (8.8), (12.22, 12.23, 12.24), (20.39), and which represent a generalization of these equations. They satisfy the general postulates of the quantum theoretical formalism, since we derive from them the validity of the operator equation $\dot{\varphi} = i/h \cdot [H, \varphi]$ for each field quantity (which does not contain the time explicitly). The corpuscular properties can be inferred again from the fact that, for instance, the eigen values of the total energy are the sums of the energies $h\omega_k$ of individual particles.

To illustrate this we state the commutation rules of Fierz for the case $s = 2$, $\mu \neq 0$ (complex field). With the same notations as above and with the abbreviation:

$$d_{\nu\nu'} = \delta_{\nu\nu'} - \frac{1}{\mu^2} \frac{\partial^2}{\partial x_\nu \partial x_{\nu'}}$$

[cf. (12.24)] they may be written:

$$\left[\psi_{\mu\nu}(x, t), \psi_{\mu'\nu'}(x', t')\right] = \left[\overset{*}{\psi}_{\mu\nu}(x, t), \overset{*}{\psi}_{\mu'\nu'}(x', t')\right] = 0,$$

$$\left[\psi_{\mu\nu}(x, t), \overset{*}{\psi}_{\mu'\nu'}(x', t')\right] = \frac{h}{i}\, c^2 \cdot \frac{1}{2}\Big(d_{\mu\mu'}\, d_{\nu\nu'} + d_{u\nu'}\, d_{\mu'\nu} - \frac{2}{3}\, d_{\mu\nu} d_{\mu'\nu'}\Big) D(x - x', t - t'). \tag{22.13}$$

One can easily verify that these relations are compatible with the field equations (22.1, 22.2, 22.3) due to the fact that the invariant D-function satisfies the Schrödinger-Gordon equation. For instance, the contraction of the right side of (22.13) with respect to the indices μ, ν is zero in accordance with (22.1), for:

$$\sum_{\nu} d_{\nu\mu'} d_{\nu\nu'} D = \left\{ d_{\mu'\nu'} - \frac{1}{\mu^2} \frac{\partial^2}{\partial x_{\mu'} \partial x_{\nu'}} \left(1 - \frac{\Box}{\mu^2}\right) \right\} D = d_{\mu'\nu'} D,$$

$$\sum_{\nu} d_{\nu\nu} D = \left(4 - \frac{\Box}{\mu^2}\right) D = 3 D.$$

We have already pointed out the importance of the fact that in the non-quantized theory the sign of the energy for non-integral spin is not definite. In order to secure a positive-definite energy in quantum theory, the most obvious procedure—and evidently also the only possible one—is to introduce the hole theory of the positron and the subtraction formalism, discussed in §20, into the theories of particles with higher non-integral spin. In particular, the vacuum must be identified with that state of the system in which all individual states of negative energy are occupied, and all states of positive energy are empty. Speaking of "occupied" (i.e., completely occupied) individual states, implies of course that the theory is quantized according to the Pauli exclusion principle; i.e., the commutation rules must as in (20.39), for instance, refer to the anticommutators (bracket symbols with positive sign). Hence the postulate that the energy in the quantized theory must be positive can be satisfied only by assuming that the particles with non-integral spin generally obey the Pauli exclusion principle.

If one tries, on the other hand, to apply the exclusion principle to particles with integral spin ($s = 0, 1, \ldots$), i.e., to require for the respective fields commutation rules with positive "bracket symbols," we arrive at a mathematical contradiction. This is due to the fact that the anticommutator of an operator with its Hermitian conjugate ($[a, a^*]_+ = aa^* + a^*a$) is positive-definite, while the expression to which the anticommutator should be equal can have both signs. Hence, for integral spin, quantization according to the exclusion principle is impossible. On the other hand, the commutation rules for commutators ($[a, a^*] = aa^* - a^*a$ is, of course, indefinite) are consistent.

The most general proof for these statements has been given by Pauli.[1] The special form of the field equations is left entirely arbitrary; moreover, no unique value for the spin must be assumed. Whether a field describes particles with integral or non-integral spin, can be seen from the transformation properties of its components under Lorentz transformations. One can now associate, in the unquantized theory, with each solution of the field equations[2] another

[1] *Phys. Rev. 58*, 716, 1940.

[2] Only their invariance under the "proper" Lorentz group is essential, i.e., transformations with the determinant +1, which do not reverse the direction of time.

solution with the help of the transformation $x_\nu \rightarrow - x_\nu$ and the simultaneous reversal of the signs for certain field components. The energy-momentum tensor which is some quadratic or bilinear function of the field functions has then the property that the values of its components for the two solutions are either of the same or of opposite sign according to whether the spin is assumed integral or half-integral. For half-integral spin the indefinite character of the energy $\left(-\int dx\, T_{44}\right)$ in the unquantized theory is thus generally proven. From it follows the necessity of quantization according to the exclusion principle. For integral spin, correspondingly, the charge is indefinite. With regard to the invariant commutation rules Pauli requires only that the commutators, or the anticommutators, of the field components be represented by the invariant D-function (4.25). Formally it would be possible here to admit the other invariant function, which was called D' in §20 [cf. (20.56)]; but since this function is not zero in the outside part of the light cone [in contrast to the D-function; cf. (4.32)], this possibility can be excluded for physical reasons: commutation rules, which contain the D'-function, would mean that measurements in world points situated spacelike to each other would not be independent. This would imply a propagation of perturbations with velocities greater than that of light (cf. §4, last part). If one excludes this, only the D-function remains an admissible invariant function. In that case it can be shown that the commutation rules with + bracket symbols can be made consistent only for half-integral spin; for integral spin such rules would lead to equations of the type $[a, a^*]_+ = 0$, which are contradictory on account of the positive-definite character of the left side, thus proving again, and most generally, that the exclusion principle cannot be applied to particles with integral spin.

The conclusion is that the relativistic quantum theory of fields compels us to quantize a field by means of commutation rules with negative or positive bracket symbols, according to whether the corresponding particles have integral or half-integral spin. This is equivalent to the statement that particles with integral spin necessarily obey Bose-Einstein statistics and those with half-integral spin necessarily the Fermi-Dirac statistics, and is just what we know from the experimental evidence to be true, at least for those elementary particles which are sufficiently well-known experimentally: light quanta, electrons, protons, and neutrons. It is certainly one of the most beautiful successes of the quantum theory of fields that—together with the postulates of the theory of relativity—it furnishes a general theoretical foundation for the connection between spin and statistics.

§ 23. Outlook

If one tries to summarize the accomplishments of the theories that have been discussed, one must distinguish between theories which involve "interactions" and those without interactions. The theory of the force-free fields, or particles, as such leaves nothing to be desired. It reconciles the descriptions in terms of waves and corpuscles in a satisfactory way. We emphasize especially how the quantumlike nature of energy, momentum, and electric charge appear in this theory as a consequence of the field quantization; furthermore, the theory leads to a natural classification of the elementary particles according to their spin values; the characteristic connection between spin and statistics can be deduced from the properties of the respective fields; the number of known simple field types is just sufficient to fit all known elementary particles.

As to the theories involving interactions, the outlook is much less satisfactory. One starts with the assumption that there exist different fields, or particles, which are coupled with each other by suitable invariant terms added to the Lagrangian. The formalism of quantum theory applied to such systems leads to some very reasonable results, for instance, regarding the forces transmitted by the fields. But in the self-energy problems one meets again and again with the divergence difficulties. The primitive "cut off" methods are at variance with the relativistic requirements. Even if an invariant subtraction formalism can be found which may succeed in eliminating all infinities from the theory, it will presumably involve a great deal of arbitrariness. Although it seems likely that the interaction theory is not altogether wrong, in particular in those problems where the special choice of the "cutting off" procedure (form factors) is irrelevant, such a theory can hardly be accepted as final or satisfactory.

We know from the classical Lorentz theory of the electron that the self-energy problem is intimately connected with the question of the "electromagnetic mass." In the classical theory, the problem was essentially that of the structure of the electron,[1] and even today, in quantum theory, this same problem is still at the root of all self-energy difficulties. Apparently it is not sufficient to assign to the electron a spatial extension, i.e., a form factor. Furthermore, this would hardly be compatible with the idea that the electron is an indivisible unit. Although the correct formulation is still unknown, the impression remains that the self-energy problem is, in fact, intimately connected with the question of the masses of the elementary particles. While

[1] Cf. Pauli, *Encykl. d. Math. Wissensch.* Vol. V, part 2, sections 63 to 67.

in the formalism so far considered, the constant μ of the rest mass plays the part of an arbitrary parameter, we encounter here the problem of the values of the masses or of the mass ratios of the elementary particles regardless of whether the interaction with other fields causes an additional inertia or not.

In the above-mentioned classical investigations, the electromagnetic field was considered the only agency (besides gravitation) responsible for the interaction so that it seemed justifiable to search for a "unitary theory" with purely electromagnetic basis. Such a limitation of our problem would no longer be suitable in the light of our present knowledge. Besides the electromagnetic forces, nuclear forces, for instance, must be considered. They must be considered almost certainly of non-electromagnetic origin even if one doubts the meson theory. One could, for instance, suppose that the masses of the proton and of the neutron are mainly determined by the inertia of the adherent "nuclear field," while the electromagnetic field would cause only small corrections. This would explain why the proton and neutron masses are approximately equal and large compared with the mass of the electron.

Besides the masses, one should also consider other characteristic properties of the elementary particles. With regard to the spin, we have seen in §22 that small spin values ($s = 0$, 1, and $\frac{1}{2}$) are distinguished by greater simplicity of the formalism. This is, however, not sufficient to explain why only particles with small spin exist in nature, let alone why certain spin values exist only combined with definite charge and mass numbers. We know just as little about the reasons why certain types of interactions between elementary particles are distinguished before others which are formally possible. This question is related to the question about the numerical value of the dimensionless constant e^2/hc, which determines the strength of the interaction between electron and light quantum[1] (Sommerfeld's fine structure constant); in quantum electrodynamics this numerical value is taken from experiments ($e^2/hc \cong 1/137$, if the elementary charge e is measured in ordinary and not in Heaviside units). Such a coupling parameter appears, of course, in any interaction term. A special case is the one in which the coupling parameter disappears, which means that the respective type of interaction does not exist. Such statements cannot be derived from the basic postulates of the theory but rather must be taken from the experiments. We may, for instance, imagine an interaction between protons, positrons, and electromagnetic field in such a way that there

[1] In the interaction terms (17.14) and (21.17), the strength of the coupling is determined by the factor e (e_n or ϵh), so that the expression $e/\sqrt{hc}$ plays the role of a dimensionless coupling parameter.

exists a spontaneous transformation of a proton into a positron with emission of a light quantum[1]; such an interaction which is formally possible must be excluded in view of the stability of the free proton.

All this indicates that the quantum theory of the wave fields in its present form is a too general frame which comprises many more theoretical possibilities than are realized in nature. Considering that the problem of the numerical values of the coupling parameters is again related to the mass problem, one must conclude that the self-energy problem can be solved only together with the whole complex of problems mentioned above. This solution seems to require an entirely new idea, for which the present theories offer not even a starting point. Until then one must be content with provisional remedies, as, for instance, the "cutting off method." This seems justifiable in view of the partial success of the classical electron theory. In this case the most beautiful and permanent results were gained by postponing the deeper questions relating to the structure and the mass of the electron. The present quantum theory of fields can be regarded as a finished discipline in the same sense as this classical theory.

It is known that one obtains for the "classical radius of the electron" the order of magnitude $e^2/mc^2 \cong 2.8 \cdot 10^{-13}$ cm., if one assumes that the electromagnetic mass is of the same order of magnitude as the actual mass m of the electron. Everything indicates that in the quantum theory, too, the "cutting off" radius does not exceed the classical radius of the electron, i.e., that only momenta $\gtrsim h \cdot mc^2/e^2 \cong 137\, mc$ are affected by the cutting off of the momentum spectrum. Accordingly, the theory should be reliable for all events in which only wave lengths $> e^2/mc^2$, i.e., momenta $< 137\, mc$, play an essential part. For certain problems the theory stands the test even in the region of higher energies; the quantum electrodynamical formulas for the probability of certain radiation processes ("Bremsstrahlung," pair production), for instance, are even at very high energies quite compatible with the observations on the cosmic

[1] Such an interaction term may be written:

$$\sum_{\varrho, \sigma, \nu} c_{\varrho\sigma\nu} \Phi_\varrho \varphi_\sigma^* \psi_\nu + \text{conj.},$$

where Φ_ρ, φ_σ, ψ_ν represent the wave functions of the proton, of the positron, and of the light field. The coefficients $c_{\rho\sigma\nu}$ are determined (except for a common factor) by the Lorentz invariance.

radiation. Heisenberg[1] has tried to define the limits of validity of the theory more accurately on the basis of plausible considerations; he introduces the conception of a "universal length," which he assumes to be of the order of magnitude of the classical radius of the electron. The region beyond these limits is unknown territory and one may hope that its experimental exploration by the study of cosmic radiation or other fundamental phenomena might eventually indicate the direction of further progress in these, as yet, unsolved problems.

---

[1] *Z. Phys.* *110*, 251, 1938.

Appendix I

The Symmetrization of the Canonical Energy-Momentum Tensor

The energy-momentum tensor $T_{\mu\nu}$ as defined in (2.8) is in general not symmetrical. It is possible to give a general formula for a tensor $\Theta_{\mu\nu} = T_{\mu\nu} + T'_{\mu\nu}$ by adding a tensor $T'_{\mu\nu}$ to the canonical $T_{\mu\nu}$ which satisfies the following three conditions:

$$\Theta_{\mu\nu} = \Theta_{\nu\mu}, \tag{I.1}$$

$$\sum_{\mu=1}^{4} \frac{\partial T'_{\mu\nu}}{\partial x_\mu} = 0, \tag{I.2}$$

$$\int T'_{4\mu}\, dx = 0. \tag{I.3}$$

The second condition ensures a differential conservation law of energy and momentum:

$$\sum_{\mu=1}^{4} \frac{\partial \Theta_{\mu\nu}}{\partial x_\mu} = 0 \tag{I.4}$$

[provided that $\partial L/\partial(x_\nu) = 0$; cf. (2.11)]
and the third condition implies that the total energy and momentum is the same for the canonical as for the symmetrical tensor:

$$\int \Theta_{4\mu}\, dx = \int T_{4\mu}\, dx. \tag{I.5}$$

The proof of this theorem follows from the relativistic invariance of the Lagrange density function L.

Let an infinitesimal Lorentz transformation be given by:

$$\delta x_\mu = \sum_{\nu=1}^{4} \omega_{\mu\nu}\, x_\nu,$$

where the infinitesimals $\omega_{\mu\nu}$ form an antisymmetrical tensor $\omega_{\mu\nu} = -\omega_{\nu\mu}$. The variation of the field variables at a fixed space-time point are then given by:

$$\text{(I.6)} \qquad \delta\psi_\sigma = \sum_\rho \Lambda_{\sigma\rho}\, \psi_\rho.$$

The $\Lambda_{\sigma\rho}$ are linear functions of the parameters $\omega_{\mu\nu}$ which characterize a given transformation, thus:

$$\text{(I.7)} \qquad \Lambda_{\sigma\rho} = \sum_{\lambda,\mu=1}^{4} \Lambda_{\sigma\rho}^{\lambda\mu}\, \omega_{\lambda\mu},$$

where we define the quantities $\Lambda_{\sigma\rho}^{\lambda\mu}$ by:

$$\Lambda_{\sigma\rho}^{\lambda\mu} = \left(\frac{\partial \Lambda_{\sigma\rho}}{\partial \omega_{\lambda\mu}}\right)_0 = -\Lambda_{\sigma\rho}^{\mu\lambda}$$

For the derivatives of the field variables we have a variation given by:

$$\text{(I.8)} \qquad \delta \frac{\partial\psi_\sigma}{\partial x_\mu} = \sum_\rho \Lambda_{\sigma\rho} \frac{\partial \psi_\rho}{\partial x_\mu} + \sum_{\nu=1}^{4} \omega_{\mu\nu} \frac{\partial \psi_\sigma}{\partial x_\nu}.$$

The variation of the Lagrangian L at a fixed space-time point is zero because L is a scalar:

$$\text{(I.9)} \qquad 0 = \delta L = \sum \frac{\partial L}{\partial \psi_\rho}\, \delta\psi_\rho + \sum_\rho \sum_{\mu=1}^{4} \frac{\partial L}{\partial \dfrac{\partial \psi_\rho}{\partial x_\mu}}\, \delta\left(\frac{\partial \psi_\rho}{\partial x_\mu}\right).$$

Inserting the equations (I.6, I.7, I.8) into (I.9) we find:

$$0 = \sum_{\lambda,\mu=1}^{4} \left\{ \sum_{\rho,\sigma} \frac{\partial L}{\partial \psi_\rho} \Lambda_{\rho\sigma}^{\lambda\mu}\, \psi_\sigma + \sum_{\nu,\rho,\sigma} \frac{\partial L}{\partial \dfrac{\partial \psi_\rho}{\partial x_\nu}} \left(\Lambda_{\rho\sigma}^{\lambda\mu} \frac{\partial \psi_\sigma}{\partial x_\nu} + \frac{\partial \psi_\rho}{\partial x_\mu}\, \delta_{\lambda\nu} \right) \right\} \omega_{\lambda\mu}.$$

It follows that the tensor:

$$\text{(I.10)} \qquad \Gamma_{\lambda\mu} = \sum_{\rho,\sigma} \left(\frac{\partial L}{\partial \psi_\rho}\, \psi_\sigma + \sum_{\nu=1}^{4} \frac{\partial L}{\partial \dfrac{\partial \psi_\rho}{\partial x_\nu}} \frac{\partial \psi_\sigma}{\partial x_\nu} \right) \Lambda_{\rho\sigma}^{\lambda\mu} + \sum_\rho \frac{\partial L}{\partial \dfrac{\partial \psi_\rho}{\partial x_\lambda}} \frac{\partial \psi_\rho}{\partial x_\mu}$$

is symmetrical:

$$\Gamma_{\lambda\mu} = \Gamma_{\mu\lambda}. \tag{I.11}$$

If we define a tensor $H_{\nu\lambda\mu}$ of third rank by:

$$H_{\nu\lambda\mu} = \sum_{\rho,\sigma} \frac{\partial L}{\partial \frac{\partial \psi_\rho}{\partial x_\nu}} \Lambda^{\lambda\mu}_{\rho\sigma} \psi_\sigma = -H_{\nu\mu\lambda} \tag{I.12}$$

then we have from the field equations (2.9):

$$\sum_{\mu=1}^{4} \frac{\partial H_{\nu\lambda\mu}}{\partial x_\nu} = \sum_{\rho,\sigma}\left(\frac{\partial L}{\partial \psi_\rho}\psi_\sigma + \sum_{\mu=1}^{4} \frac{\partial L}{\partial \frac{\partial \psi_\rho}{\partial x_\nu}} \frac{\partial \psi_\sigma}{\partial x_\nu}\right)\Lambda^{\lambda\mu}_{\rho\sigma} \tag{I.13}$$

Equation (I.10) may then be written from (2.8):

$$\Gamma_{\lambda\mu} = \sum_{\nu=1}^{4} \frac{\partial H_{\nu\lambda\mu}}{\partial x_\nu} - T_{\lambda\mu} + L\,\delta_{\lambda\mu}. \tag{I.14}$$

We construct now a tensor $G_{\nu\lambda\mu}$ which satisfies the two conditions:

$$G_{\nu\lambda\mu} = -G_{\lambda\nu\mu} \tag{I.15}$$

$$1/2\,(G_{\nu\lambda\mu} - G_{\nu\mu\lambda}) = H_{\nu\lambda\mu} \tag{I.16}$$

The tensor G is uniquely defined by these conditions and is given by:

$$G_{\nu\lambda\mu} = H_{\mu\lambda\nu} + H_{\lambda\mu\nu} + H_{\nu\lambda\mu}. \tag{I.17}$$

The tensor:

$$\Theta_{\lambda\mu} = T_{\lambda\mu} + T'_{\lambda\mu} \tag{I.18}$$

(I.19) with:

$$T'_{\lambda\mu} = -\sum_{\nu=1}^{4} \frac{\partial G_{\nu\lambda\mu}}{\partial x_\nu}$$

satisfies then all the conditions of the symmetrical energy-momentum tensor:

$$\Theta_{\lambda\mu} - \Theta_{\mu\lambda} = T_{\lambda\mu} - T_{\mu\lambda} - 2\sum \frac{\partial H_{\nu\lambda\mu}}{\partial x_\nu} = -\Gamma_{\lambda\mu} + \Gamma_{\mu\lambda} = 0 \tag{I.1}$$

$$\sum_{\lambda=1}^{4} \frac{\partial T'_{\lambda\mu}}{\partial x_\lambda} = -\sum_{\nu,\lambda=1}^{4} \frac{\partial^2 G_{\nu\lambda\mu}}{\partial x_\nu\,\partial x_\lambda} = 0 \tag{I.2}$$

on account of (I.15).

$$\text{(I.3)} \qquad \int T'_{4\mu}\,dx = -\int\sum_{\nu=1}^{4}\frac{\partial G_{\nu4\mu}}{\partial x_\nu}\,dx = -\int\sum_{k=1}^{3}\frac{\partial G_{k4\mu}}{\partial x_k}\,dx = 0.$$

It is easy to verify and may be left to the reader that the tensors defined in (12.50) and (20.10) for vector and spinor fields, respectively, are precisely the tensors obtained with the general rule (I.18, I.19). This rule is completely general and allows the definition of a symmetrical energy-momentum tensor for any kind of a field with an invariant Lagrange function.

INDEX

A

Alvarez, 110
Angular momentum, of a field, 10
  of particles. See *Spin*.
Anticommutator, 174
Arnold, W. R., 110

B

Bartlett force, 105
Belinfante, 10
Bethe, 108, 196
Bhabha, 97
Bloch, 110, 140, 145
Bohr, 26, 115
Booth, 97
Born, 46, 71, 201
Bose statistics for light quanta, 123
Bose-Einstein statistics, impossibility for particles with half odd integral spin, 211
  quantization according to, 36 ff.
Breit, 105, 107, 110

C

Canonical equations of motion as operator equations, 7
Canonical field equations, 7
Canonically conjugate field, 3 ff., 12
Charge, charge density, ambiguity in their definition, 208
Charge, charge density (electric), 14 ff.
Charge eigenvalues as integral multiples of an elementary charge, 51 ff.
Charge independent forces, 66
Charge number, 54
Charge symmetry in the theory of the positron, 186 ff.
Christy, 97
Commutation relations, invariant, 20 ff., 24
  canonical, 4 ff.
  for non-simultaneous quantities, 17 ff.
  for the electromagnetic field, 115
  for exclusion particles, 174 ff.
Commutation relations (*continued*):
  in momentum space, 27 ff.
  in the theory of particles with higher spin, 209
Complex fields, 12 ff.
Condon, 105
Conservation laws for energy, momentum and angular momentum, 8 ff.
  for the charge, 14 ff.
Continuity equation(s), for energy and momentum, 8 ff.
  as operator equations, 11, 15
  for the charge, 14 ff.
Corben, 95
Coulomb potential, 132, 150, 158 ff., 193
  scattering of mesons, 71, 96 ff.
Coupling of mesons with nucleons (nuclear) 37 ff., 53 ff., 97 ff.
Coupling of mesons with nucleons, (strong), 63
Current density (electric), 14 ff.
Cutting-off method, 46, 212

D

D-function, invariant, 23 ff.
  of Jordan and Pauli, 115, 148 ff.
D′-function, invariant, 185
Delbrück, 200
Delta-function (Dirac's), 5
  Fourier expansion, 23
Derivative, functional, 2
  time, of an operator, 7
Deuteron, electric quadrupole moment, 108
  stationary states, 106 ff.
Dirac, 5, 25, 112, 127 ff., 135, 138, 153 ff., 167, 168 ff., 180, 183 196 ff., 203
Dirac's radiation theory, 135 ff.
Dirac's theory of the positron (theory of holes), 180 ff.
Dirac's wave equation of the electron, 168, 188

E

Einstein, 206 ff.
Electric charge and current density, 14
Electromagnetic field, 111 ff.
  interaction with electrons, 124 ff., 192 ff.
  interaction with scalar mesons, 126
  quantization, 114 ff.
Electromagnetic mass, 153, 165 ff., 212 ff.
Electron radius, classical, significance for limits of validity of the theory, 214 ff.
Electron wave field, 167 ff.
Electron wave field, quantization according to the exclusion principle, 172 ff.
Energy, conservation of, 8
Energy current, 9
Energy density, 8
  positive definite character, 208
Energy eigenvalues as sums of corpuscle energies, 35
Energy-momentum tensor, 9
  ambiguity, 208
  symmetrical, construction, 10, 217
Euler, 201
Exchange force, exchange operator, 60 ff., 100 ff.
Exclusion principle, impossibility for particles with integral spin, 210
  quantization according to, 172 ff.

F

Feenberg, 107
Fermi, 112, 135 ff.
Fermi-Dirac statistics, impossibility for particles with integral spin, 210
  quantization according to, 172 ff.
Field energy, 8
Field energy, positive definite character, 208
Field equations, as operator equations, 7
  canonical, 7
  canonically conjugate, 3 ff., 12
  classical, 1 ff.
Field functions, 1
  complex, 12
Field operators, 5
  matrix representation, 34 ff., 51 ff., 171 ff.
Field operators (*continued*):
  non-Hermitian, 13
  time dependent, 17 ff.
Fierz, 203 ff.
Fine structure constant, 213 ff.
Fock, 112, 138, 183
Form-factor of the nucleon, 46, 62
Fourier decomposition of a field, 27 ff.
Frisch, 110
Fröhlich, 97, 109
Functional derivative, 2
Furry, 183

G

Gardner, 110
Gauge transformation, first kind, 14
  second kind, 68, 111 ff., 189, 205, 207
Gordon, 67
Gravitational waves, 207

H

Hafstad, 105
Halpern, 200
Hamiltonian, as energy density, 8 ff.
  of a field, 4
Hamiltonian operator, 6
Heavy meson, 110
Heisenberg, 4 ff., 18, 54, 101, 105, 112, 135, 183, 196 ff., 215
Heisenberg force, 101, 105
Heitler, 26, 97, 109, 136, 196
Heydenburg, 105
Hilbert, 207
Hoisington, 107, 110
Hole theory of the positron, 180 ff., 210
Hu, 109
Hyper quantization, 168

I

Indeterminacy relations for field components, 26, 115
Interaction, nuclear. See *Coupling*.
Invariant D-functions, 23 ff.
Isobars of the proton, 63
Isotopic spin, matrices with respect to the isotopic spin, 57
Iwanenko, 105

J

Jauch, 109, 208
Jordan, 24, 115, 148 ff., 172

K

Kellogg, 108, 110
Kemmer, 63 ff., 97, 104, 107, 109, 201
Klein-Nishina formula, 196
Koba, 193
Kobayasi, 97
Kramers, 136
Kusaka, 97
Kusch, 110

L

Lagrangian of a field, 3
  Lorentz invariance, 15 ff.
  transformation, 2, 208
Laporte, 97
Lattes, 110
Light quantum, 122 ff.
  spin, 123
Limits of validity of the theory, 214
Longitudinal waves, electromagnetic, elimination, 120 ff., 130 ff., 156 ff., 193
Lorentz invariance, of the commutation relations, 17 ff.
  of the Lagrangian, 15
Lorentz invariant functions, 23 ff.

M

Magnetic moment, 208
  of proton, neutron, and deuteron, 109 ff.
  of the vector meson, 94 ff.
Majorana, 101
Majorana force, 101, 105
Mass, electromagnetic, 153, 165 ff., 212 ff.
Matrices, with respect to particle numbers, 34, 51 ff., 172 ff.
  with respect to the charge number, 55 ff.
  with respect to the spin coordinates, 127, 168 ff.
Matrix representation of field operators, 34 ff., 51 ff., 171 ff.
Maxwell's equations, as operator equations, 117 ff., 129 ff.
  for the expectation values of the field-strengths, 118, 129 ff.
Meson, mesotron, 37
  charged, in an electromagnetic field, 66 ff., 90 ff., 126
Meson absorption and emission, 41, 54 ff.
Meson field, 37
  charged, 48 ff.
  pseudoscalar, 109 ff.
  vector, 75 ff.
Meson scattering, electric, 71, 95 ff.
  nuclear, 42 ff., 58 ff., 100
Mie, 201
Millman, 110
Møller, 109
Møller-Rosenfeld coupling in meson theory, 109
Momentum of a field, 10
Momentum density, 10
Momentum eigenvalues as sums of corpuscle momenta, 35
Momentum space, transition to, 27 ff.
Muirhead, 110
Multi-body forces, 63
Multiple time formalism of the electromagnetic field, 138
  physical interpretation of the multiple-time Schrödinger function, 145

N

Nishina, 196
Nuclear forces, due to meson fields, 43 ff., 59., 99 ff.
  phenomenological potential, 105 ff.
  range, 45, 110
  saturation conditions, 107
Nuclear interaction. See *Coupling.*
Nucleon, 54

O

Occhialini, 110
Operators, time dependent, 16 ff.
Oppenheimer, 63, 183, 196

P

Pair production, annihilation of mesons, 70, 73, 97
Pair production (annihilation), electrons and positrons, 194, 196
  mesons, 70, 73, 97
Pauli, 4 ff., 14, 18, 24, 48, 57, 68, 72, 95, 102, 109, 112, 115, 135, 148 ff., 166, 167, 187, 189, 205, 207, 210 ff., 212
Peierls, 183

Photon. See *Light quantum.*
Plesset, 196
Podolsky, 112, 138
Polarization of the vacuum, 73, 196, 200
Positron, theory of holes, 180 ff.
Powell, 110
Present, 105
Proca, 75
Proton-isobars, 63
Proton-neutron, 54
Pseudoscalar field, 109 ff.

Q

Quadrupole moment of the deuteron, 108
Quantization of a field
  according to Bose-Einstein statistics, 36 ff.
  according to Pauli's exclusion principle or Fermi-Dirac statistics, 172 ff.
  canonical, 5
  second, 168
Quantum electrodynamics, 111 ff.
  relativistic and gauge invariance, 145

R

Rabi, 108, 110
Ramsey, 108, 110
Range of nuclear forces, 45, 110
Rarita, 110
Reality conditions for Fourier coefficients, 27
Relativistic invariance. See *Lorentz invariance.*
Roberts, 110
Rosenfeld, 10, 18, 26, 109, 115

S

Sakata, 97
Saturation conditions for the nuclear forces, 107
Sauter, 196
Scalar field, complex, 48 ff.
  real, 29 ff.
Scattering, electric, of mesons, 71, 95 ff.
  nuclear, of mesons, 42 ff. 58 ff., 100
  of light by light or electric fields, 200 ff.
    by electrons, 137 ff., 194 ff.
Schrödinger-Gordon equation, 22, 67
Schwinger, 63, 95, 109, 110
Self-energy, 46, 60, 73, 100, 133, 151 ff., 201, 212 ff.
Share, 105, 110
Sommerfeld's fine structure constant, 213 ff.
Spin, of electron and positron, 183
  of the light quantum, 123
  of the vector meson, 89 ff.
  particles with higher spin, 203 ff.
  relation between spin and statistics, 210 ff.
Stern, 110
Stress tensor, 9
Stueckelberg, 39, 97
Subtraction prescriptions, 180 ff., 196 ff.
Symmetrical theory for the meson field, 63 ff., 104

T

Taketani, 97
Tamm, 105
Tati, 193
Time dependent operators, 16 ff.
Tomonaga, 193
Tuve, 105

U

Unitary theory, 201, 213
Universal length, 215
Utiyama, 97

V

Vacuum polarization, 73, 196, 200
Van der Waerden, 203
Variational principle, definition of the field equations, 1 ff., 205
Vector meson fields, 75 ff.

W

Waller, 164
Weisskopf, 48, 72, 73, 201
Wentzel, 63, 109, 136, 151, 153
Wigner, 105, 172
Wigner force, 105
Wilson, 97

Y

Yukawa, 45, 53, 59 ff., 97, 104
Yukawa potential, 44, 106
Yukawa's theory of nuclear forces, 45, 59 ff., 99 ff.

Z

Zacharias, 108, 110